Lecture Notes in Mechanical Engineering

Lecture Notes in Mechanical Engineering (LNME) publishes the latest developments in Mechanical Engineering—quickly, informally and with high quality. Original research reported in proceedings and post-proceedings represents the core of LNME. Also considered for publication are monographs, contributed volumes and lecture notes of exceptionally high quality and interest. Volumes published in LNME embrace all aspects, subfields and new challenges of mechanical engineering. Topics in the series include:

- Engineering Design
- Machinery and Machine Elements
- Mechanical Structures and Stress Analysis
- Automotive Engineering
- Engine Technology
- Aerospace Technology and Astronautics
- Nanotechnology and Microengineering
- Control, Robotics, Mechatronics
- MEMS
- Theoretical and Applied Mechanics
- Dynamical Systems, Control
- Fluid Mechanics
- Engineering Thermodynamics, Heat and Mass Transfer
- Manufacturing
- Precision Engineering, Instrumentation, Measurement
- Materials Engineering
- Tribology and Surface Technology

More information about this series at http://www.springer.com/series/11236

Alexander N. Evgrafov
Editor

Advances in Mechanical Engineering

Selected Contributions from the Conference
"Modern Engineering: Science
and Education", Saint Petersburg, Russia,
June 2017

 Springer

Editor
Alexander N. Evgrafov
Peter the Great St. Petersburg Polytechnic
   University
Saint Petersburg
Russia

ISSN 2195-4356   ISSN 2195-4364   (electronic)
Lecture Notes in Mechanical Engineering
ISBN 978-3-319-72928-2   ISBN 978-3-319-72929-9   (eBook)
https://doi.org/10.1007/978-3-319-72929-9

Library of Congress Control Number: 2017931350

Printed on acid-free paper

This Springer imprint is published by Springer Nature
The registered company is Springer International Publishing AG
The registered company address is: Gewerbestrasse 11, 6330 Cham, Switzerland

Preface

The "Modern Mechanical Engineering: Science and Education" (MMESE) conference was initially organized by the Mechanical Engineering Department of Peter the Great St. Petersburg Polytechnic University in June 2011 in St. Petersburg, Russia. It was envisioned as a forum to bring together scientists, university professors, graduate students, and mechanical engineers, presenting new science, technology, and engineering ideas and achievements.

The idea of holding such a forum proved to be highly relevant. Moreover, both the location and timing of the conference were quite appealing. Late June is a wonderful and romantic season in St. Petersburg—one of the most beautiful cities, located on the Neva river banks and surrounded by charming greenbelts. The conference attracted many participants, working in various fields of engineering: design, mechanics, materials, etc. The success of the conference inspired the organizers to turn the conference into an annual event.

More than 80 papers were presented at the sixth conference MMESE-2017. They covered topics ranging from the mechanics of machines, material engineering, structural strength, and tribological behavior to transport technologies, machinery quality, and innovations, in addition to dynamics of machines, walking mechanisms, and computational methods. All presenters contributed greatly to the success of the conference. However, for the purposes of this book, only 20 papers, authored by research groups representing various universities and institutes, were selected for inclusion. I am particularly grateful to the authors for their contributions and all the participating experts for their valuable advice. Furthermore, I thank the staff and management of the university for their cooperation and support, and especially, all members of the program committee and the organizing committee for their work in preparing and organizing the conference.

Last but not least, I thank Springer for its professional assistance and particularly Mr. Pierpaolo Riva who supported this publication.

Saint Petersburg, Russia   Alexander N. Evgrafov

Contents

Hot Orbital Forging by Tool with Variable Angle of Inclination

Leonid B. Aksenov and Sergey N. Kunkin

Abstract The paper presents a new orbital forging technology for manufacturing of axisymmetric components with massive flange parts. The tall formed part in this process leads to the buckling and formation of folds of the deformable workpieces. Two stages of orbital forging are proposed in which the tool inclination changes during processing. In the first stage, the technology of upsetting is realized without the use of the movement of precession and tool inclination. In the second stage, a cone-shaped tool corresponding to the configuration of a part is set at an angle $\gamma = 2°$ and enables the movement of precession for the upper die. Computer simulation of two stages of hot rotary forging for a part of aluminum bronze shows the possibility of forming without shape defects and metal fractures. The technology is particularly effective in small scale production of axisymmetric parts with flanges from rod workpieces.

Keywords Orbital forging · Buckling of the workpiece · Axisymmetric parts
Massive flanges · Aluminum bronze · Computer simulation

Introduction

Large quantities of parts such as flanges are used in a variety of industries. The nomenclature of these parts is very diverse and conforms to various standards, which can be determined according to individual country or Commonwealth, for example, by countries in the Common Market. Manufacture of flange parts is carried out according to different technologies, but most of them do not have a high utilization rate of metal.

L. B. Aksenov (✉) · S. N. Kunkin
Peter the Great St. Petersburg Polytechnic University, St. Petersburg, Russia
e-mail: l_axenov@mail.spbstu.ru

S. N. Kunkin
e-mail: kunkin@spbstu.ru

A. N. Evgrafov (ed.), *Advances in Mechanical Engineering*, Lecture Notes in Mechanical Engineering, https://doi.org/10.1007/978-3-319-72929-9_1

Fig. 1 Axisymmetric parts with thick flanges

In many industries, there is a large range of axisymmetric parts with massive flanges of similar configuration (Fig. 1). Alloys of non-ferrous metals, particularly bronze without tin, are often used for details of this type. These materials have a low plasticity under cold conditions and only allow for slight deformation without fracture, insufficient for shaping parts in required forms. Cast billets for the data parts can't be used because of special requirements in regard to mechanical properties and porosity.

Traditional hot die forging is not economically advantageous, because of the small series involved (200–500 pieces per year), as well as the large machining allowances that increase the complexity of their subsequent machining.

For the manufacture of such parts, the most promising process seems to be orbital forging, which in its 100-year history of development has shown its effectiveness in small series production of axisymmetric parts [1–5]. The main feature of orbital forging of massive flanges is the need to deform a considerable volume of metal to form the flange of the part. For such processes, a cylindrical workpiece with a height-to-diameter ratio of more than two is required. This leads to negative consequences, such as loss of stability of the workpiece and its subsequent crushing by the forging roll.

Control capabilities for the metal flow during orbital forging are very limited. The use of restraining rollers, flanges and so on complicate tooling to a very significant degree. It is more effective to control the direction of metal flow by changing the direction of the friction forces which act on the contact surface between the forging die and a deformable metal [6, 7]. So, to receive the outer flanges from the pipe blanks requires that the metal flow be in the direction from the billet center to its periphery. To do this, the cylindrical rollers should be placed with some offset relative to the transverse axis of the workpiece, and tapered relative to the longitude [8].

In orbital forging, it is possible to significantly affect the change in the direction of movement of the metal by changing the angle of inclination of the forging roll. This processing method was used for the orbital forging of parts made of powder materials [9]. In the first stage of the process, the seal of powder blanks was done by the upper die without a slope, and then the compact billet was finally formed by the inclined die. Similar technology was tested in orbital forging of tube-blanks by the technology of outward-flanging, in which, at the first stage, the forging die roller was used without slope for expansion of the upper part of the blank, and then the desired shape of the workpiece was formed with the inclined roll.

Computer simulation in the last 20 years has become a powerful tool for assessing new technologies and their optimization [10–13]. In this paper, based on the simulation of the process, the possibility of application of hot orbital forging to produce parts with massive flanges using two-stage technologies with different angles of inclination of the deforming tool at each stage is shown.

Research Methodology

The object of study is selected technology of hot orbital forging of billet parts with a flange-type "insert", with the 3D view shown in Fig. 2a and the main dimensions in Fig. 2b.

The technology of the production of parts was focused on the use of an orbital forging press PXW-160 [14] with a nominal force of 1600 kN. The working area of this machine is shown in Fig. 3. The parameters of the working area allow the press to form parts with a diameter up to 100–120 mm and use the billet with a height of up to 130–135 mm.

Material details—Bronze 9-4-4-1 (Russian Standard); the chemical composition is presented in Table 1. This type of bronze applies to the aluminum bronzes, which are poorly deformed in a cold state, have high strength at elevated temperatures, and have good corrosion, erosion and cavitation resistance.

The mechanical properties of the alloy Bronze 9-4-4-1 at T = 20 °C: tensile strength σ_B = 650 MPa, elongation δ = 15%. The poor plasticity in a cold state leads to the destruction of the workpieces during orbital forging [13–18].

A problem with the production of this range of components is that the use of orbital forging technology for blanks with a tall formed part leads to two types of defect: first, the buckling of the workpiece when its upper part is upset (Fig. 4a), and second, the formation of folds in the middle part of the deformable workpiece (Fig. 4b).

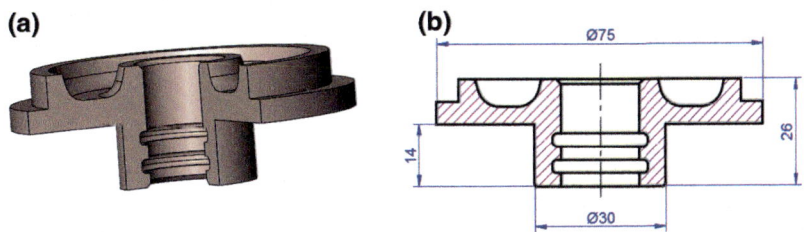

Fig. 2 Part with flange type "insert": 3D view (**a**) and principal dimensions (**b**)

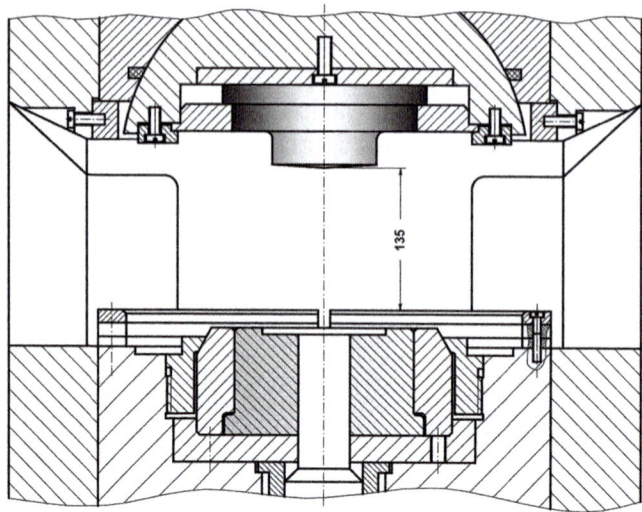

Fig. 3 The working area of the orbital forging press PXW-160

Table 1 Chemical composition of Bronze 9-4-4-1 (Russian Standard GOST 18175-78)

Marking stamps	Chemical composition (%)			
	Al	Fe	Ni	Mn
Bronze 9-4-4-1	8.8–10.0	4.0–5.0	4.0–5.0	0.5–1.2

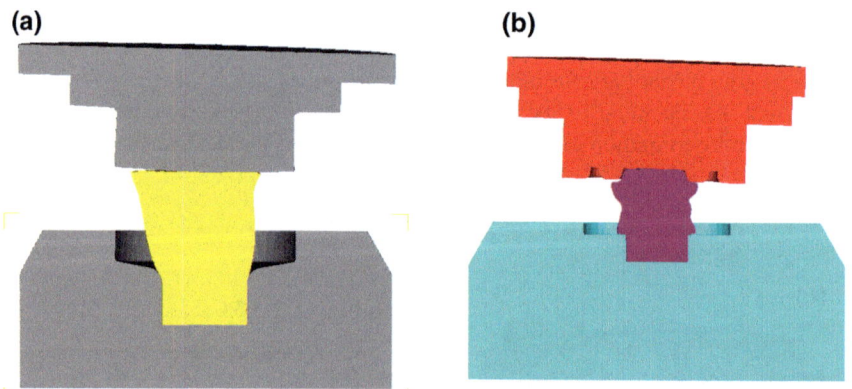

Fig. 4 Defects arising in the orbital forging of tall blanks: **a** loss of stability (buckling) in the initial stage of the process; **b** formation of folds in the middle part of the workpiece

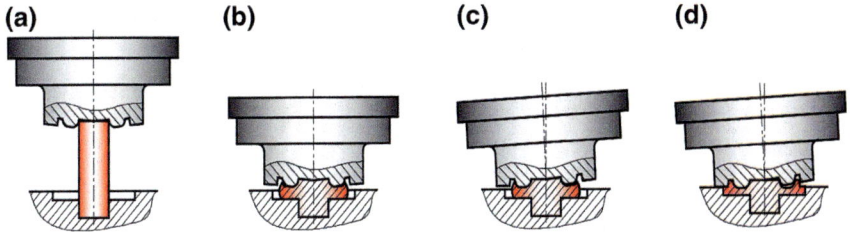

Fig. 5 Two-stage orbital forging: **a, b** first stage—upsetting without movement of precession; **c, d** second stage—orbital forging with movement of precession of the upper die

Upsetting of a cylindrical workpiece on the orbital forging press differs significantly from the traditional process of upsetting by flat plates. When the orbital forging deforming force acts on only part of the end surface of the workpiece, the load ends up being not coincident with the axis of symmetry of the cylindrical workpiece. In addition, the friction force increases the bending moment applied to the workpiece in the plane of contact of the tool with the workpiece. This significantly reduces the stability of the workpiece. So, if rods of comparable size lose stability under an axial load at a ratio of height of cylinder to diameter of 2.5 or more, under the action of the described forces, the buckling of the workpiece can occur when this ratio is less than a 2.0.

To solve these problems, a technological process of production of these parts divided into two stages is proposed (Fig. 5). The first stage is realized on an orbital press without the use of the movement of precession (Fig. 5a, b). The penetration at the stage of upsetting the conical tool that is symmetrical relative to the axis of the workpiece tool into the workpiece helps to preserve the stability of the workpiece. In the second stage (Fig. 5c, d), a cone-shaped tool corresponding to the configuration of a part is set at an angle $\gamma = 2°$ and enables the movement of precession for the upper die. Thus, for the selected equipment, the process has two stages that are discrete in nature.

Computer Simulation

Simulation of the two-stage technological process of manufacturing a part "insert" was executed through the software product Deform 3D. As a foreign counterpart of Russian Bronze 9-4-4-1, CuAl8 material was selected, with a heating temperature of the workpiece being 800 °C, and the temperature of the tool being 50 °C. In the first stage, the friction coefficient was 0.3, and at operation of orbital forging, 0.1; the number of elements was 56,000, and there was a special area with finer mesh in the volume of the deformable part of the workpiece. A cylindrical billet with a diameter of 32 mm and a height of 63.5 mm was deformed according to the scheme of upsetting, with an angle inclination $\gamma = 0°$ and an coning angle 2° of the upper

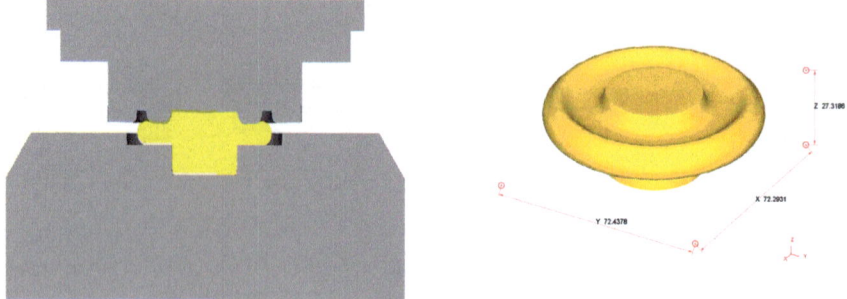

Fig. 6 Tool position and the principal dimensions of the workpiece after the stage of upsetting by the upper die without inclination

conical tool. The upsetting was carried out up to the diameter of 72.5–73.0 mm and the height of the workpiece: 27.3–28.0 mm. The force of the upsetting did not exceed 1250 kN. Buckling of the blanks was not observed, and, accordingly, did not create the conditions for further folding. After the stage of upsetting the angle of inclination of the upper die was established, $\gamma = 2°$. The process of orbital forging continued until complete formation of the flange (Fig. 6).

Discussion

Analysis at the final stage of orbital forging (Fig. 7a) of values of effective strain in the cross-section of the items received (Fig. 7b) shows that its highest value is observed in the cavity of the flange (up to 3.0). In this zone, the stress state close to the 3D state displays uneven compression. That is why the discontinuity of the metal in this area can have low probability. A relatively small effective strain is observed near the upper surface of the part. Temperature is highest (up to 4450 °C) in this area (Fig. 7c), which creates favorable conditions for forming additional elements on this surface, e.g. teeth. All this allows for considering the technique of

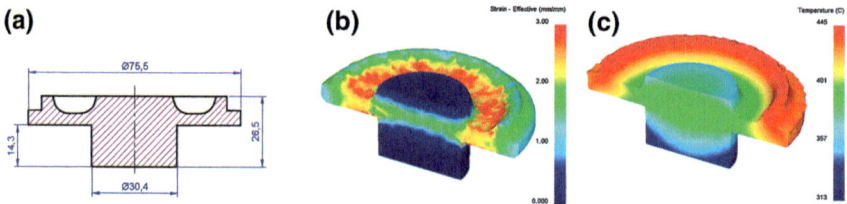

Fig. 7 Geometry of the formed part (**a**), the distribution of effective strain (**b**) and temperature (**c**) in the cross-section of details at the final stage of forming

hot rotary forging with a variable angle tool inclination as a promising one for the manufacturing of parts with massive flanges.

The present period of the development of machines for face rotary forging, which include orbital forging machines, is characterized by the creation of multi-functional machines, capable of implementing different kinematics by means of CNC. These machines are designed primarily for the processing of composites and exotic materials, including those used in the aerospace industry [19], and they can be successfully used for the implementation of orbital forging technology by changing the angle of the forming roll in the forming process.

Resume

- Use of pre-heated workpieces from aluminum bronze allows for the manufacture of blanks for parts such as massive flanges without defects of form and metal fractures in the process of forming.
- Computer simulations showed that the use of a two-stage technology helps in avoiding buckling of the workpiece in the first stage of orbital forging, and in the second (final) stage, in realizing the advantages of orbital forging to reduce the force of deformation compared to conventional upsetting.
- Modern rotary forging machines with variable kinematics and CNC systems open up broad prospects for the use of rotary forging technology with a variable angle of tool inclination.

References

1. Nowak J, Madej L, Ziolkiewicz S, Plewinski A, Grosman F, Pietrzyk M (2008) Recent development in orbital forging technology. Int J Mater Form 1 (Suppl 1):387–390
2. Han X, Hua L (2009) Comparison between cold rotary forging and conventional forging. J Mech Sci Technol 23:2668–2678
3. Aksenov LB, Kunkin SN (2016) Development of rotary forging machines: from idea to additive technologies. Sciences of Europe, Praha, Czech Republic, vol 2, # 7(7), pp 4–11
4. Extended application range for orbital forming technology. Schmid Press Release, Heinrich Schmid Machines, Tools & Dies Ltd., Jona, Switzerland, May 2005
5. Plancak ME, Vilotic DZ, Stefanovic MC, Movrin DZ, Kacmarcik IZ (2012) Orbital forging—a possible alternative for bulk metal forming. J Trends Dev Mach Assoc Technol 16(1):35–38
6. Montoya I, Santos MT, Pérez I, González B, Puigjaner JF (2008) Kinematic and sensitivity analysis of rotary forging process by means of a simulation model. Int J Mater Form 1 (Suppl 1): 383–386
7. Han XH, Hua H (2011) Effect of position between upper die and workpiece on cold rotary forging. Adv Mater Res 189–193:2547–2552
8. Aksenov LB, Kunkin SN (2015) Metal flow control at processes of cold axial rotary forging. In: Evgrafov A (ed) Advances in mechanical engineering, lecture notes in mechanical

engineering. Published by Springer International Publishing, Switzerland, pp 175–181. ISSN 2195-4356. https://doi.org/10.1007/978-3-319-29579-18

9. Standring PM (1999) The significance of nutation angle in rotary forging. In: Advanced technology of plasticity, proceeding of the 6th ICTP, vol III, pp 1739–1744
10. Deng XB, Hua L, Han XH (2011) Numerical and experimental investigation of cold rotary forging of a 20CrMnTi alloy spur bevel gear. Mater Des 32:1376–1389
11. Liu G, Yuan S, Zhang M (2001) Numerical analysis on rotary forging mechanism of a flange. J Mater Sci Technol 17(1):129–131
12. Wang GC, Zhao GQ (2002) Simulation and analysis of rotary forging a ring workpiece using finite element method. Finite Elem Anal Des 38(12):1151–1164
13. Munshi M, Shah K, Cho H, Altan T (2005) Finite element analysis of orbital forming used in spindle/inner ring assembly. In: 8th ICTP 2005—international conference on technology of plasticity, Verona, 9–13 Oct 2005
14. Kocańda A (2015) Development of orbital forging processes by using Marciniak rocking-die solutions. In: Tekkaya AE, Homberg W, Brosius A (eds) 60 excellent inventions in metal forming. Springer-Verlag, Berlin, Heidelberg, pp 319–324
15. Han X, Hua L (2013) 3D FE modelling of contact pressure response in cold rotary forging. Tribol Int 57:115–123
16. Zhuang W, Dong L (2016) Effect of key factors on cold orbital forging of a spur bevel gear. J Cent South Univ 23:277–285
17. Samołyk G (2013) Investigation of the cold orbital forging process of an AlMgSi alloy bevel gear. J Mater Process Technol 213:1692–1702
18. Han X, Hua L, Zhuang W, Zhang X (2014) Process design and control in cold rotary forging of non-rotary gear parts 214:2402–2416
19. MJC Engineering & Technology. Rotary Forging Brochure. [Ehlektronnihyj resurs]. Retrieved: http://www.mjcengineering.com/. Accessed 5 Dec 2016

Modeling and Simulation of Tapping Mode Atomic Force Microscope Through a Bond-Graph

Mohammad Reza Bahrami and A. W. Buddimal Abeygunawardana

Abstract This paper presents the bond-graph modeling of the Atomic force microscope. The Atomic force microscope is modeled as a lumped parameter system in its dynamic contact mode of operation. The Derjaguin–Muller–Toporov (DMT) force is considered as the interaction of the cantilever tip with the sample surface, and it introduces the nonlinearity to the model. The response of the model is obtained through bond graph by using a 20-sim program. Results are compared with results obtained by SIMULINK in MATLAB.

Keywords AFM · Modeling · Tapping mode · Vibration · Bond-graph 20-sim

Introduction

The atomic force microscope (AFM), developed in the mid-1980s [1], is used to magnify surface features. By using the AFM, it is possible to scan an object's surface topography with extremely high magnifications, up to $1,000,000\times$. Scanning in three dimensions is considered to be one of the most important features of an AFM, i.e., the horizontal X-Y plane and the vertical Z dimension.

A schematic of a typical AFM is illustrated in Fig. 1. A conventional AFM consists of a microcantilever with a sharp tip, a piezo scanner, and a photodetector for receiving a laser beam reflected off the end-point of the beam to provide cantilever deflection feedback.

The AFM works in three different operation modes, namely, contact, non-contact, and dynamic contact (tapping mode). The cantilever is excited at or close to its fundamental resonance frequency. Tip-sample interactions, which are

M. R. Bahrami (✉) · A. W. B. Abeygunawardana
Peter the Great Saint-Petersburg Polytechnic University, Saint Petersburg, Russia
e-mail: mr.bahrami@inbox.ru

A. W. B. Abeygunawardana
e-mail: awbuddimal@gmail.com

© Springer International Publishing AG 2018
A. N. Evgrafov (ed.), *Advances in Mechanical Engineering*, Lecture Notes in Mechanical Engineering, https://doi.org/10.1007/978-3-319-72929-9_2

9

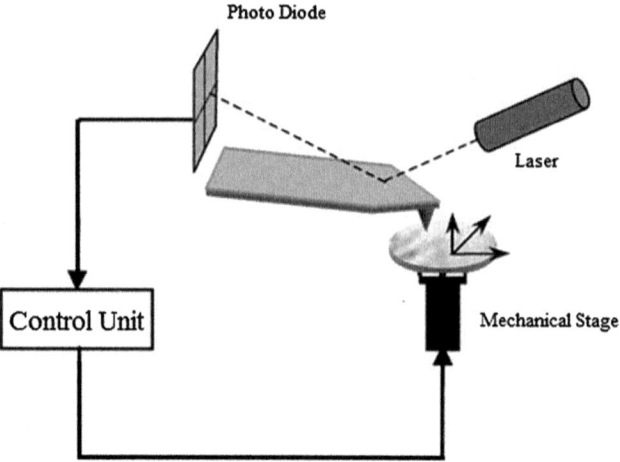

Fig. 1 Schematic of a conventional AFM

nonlinear forces, change the amplitude, phase, and frequency of oscillation. Using these changes in cantilever oscillation with respect to the external excitation plays the key role in creating sample surface topography. Therefore, the dynamic behavior of the cantilever plays a key role in a different mode of operations.

There are different methods for obtaining a mathematical model of the AFM in a different mode of operations [2–8]. Bond-graphs are one way to derive a mathematical model of systems [9–11].

Bond-graphs represent a graphical explanation of the dynamic behavior of physical systems, independent from the domain. In other words, bond-graphs describe physical systems from different domains of energy in the same way. The basic ideas of bond-graphs are energy and energy exchange.

A bond-graph divides complex systems into simple subsystems. The bond-graph gives a model of the system by analyzing these subsystems, which can be in different energy domains. Moreover, one can use this model to simulate the dynamic behavior of the system.

In general, the transfer of energy or power between subsystems is enabled by means of engineering links, e.g., mechanical shafts, electrical wires, etc. Choosing energy as the exchange variable for a model naturally leads to the use of two co-variables in each energy domain, conventionally called effort (e) and flow (f), where energy $E = \int e.f dt$. The power is a product of these two variables. The power variables (effort and flow) have different meanings in different physical domains (mechanical, electrical, hydraulic, thermal, chemical systems).

In this article, the nonlinear vibration of an AFM cantilever operating in dynamic contact mode is considered. A bond-graph is used to model the system.

Modeling of an AFM Cantilever Through a Bond-Graph

The lumped parameter model of the cantilever consists of an effective mass m_e at the end of a massless spring with the stiffness of K and a massless damper with the coefficient b, as shown in Fig. 2.

As shown in the figure, Z is the position of the cantilever tip, $F_{EXT} = F_0 \cos(\Omega t)$ is the external driving force, and $P(Z(t))$ is the force acting on the tip resulting from the tip-sample interaction. For dynamic contact AFM, a DMT sphere–plane interaction force is used as

$$P(Z(t)) = F_{DMT}(Z(t)) = -\frac{A_H R}{6a_0^2} + \frac{4}{3}E^*\sqrt{R}(a_0 - Z_0 + Z)^{3/2}, \qquad (1)$$

where A_H, R and Z_0 are the Hamaker constant, the tip apex radius, and the distance from the fixed base frame coordinate to the sample, respectively.

First, we start by explaining a systematic way to create a bond-graph. In our case, the external driving force F_{EXT} and the nonlinear tip-sample interaction force $P(Z(t))$ make the body move from its equilibrium position. They are represented by Se1 and Se2 in Fig. 3. The conversion energy law is guaranteed, since all of the power from a source of efforts goes through the other three elements of the system (Fig. 3). These elements are: capacitor or flow storage element C (spring), inertia or effort storage element I (mass) and resistor $R4$ (damper). In this case, all of the elements have the same velocity and the same flow. Flows of all bonds are identical (1-junction) in a common flow junction. 1-junctions have equality of flows, while the efforts sum up to

Fig. 2 Lumped-parameter
model of the AFM

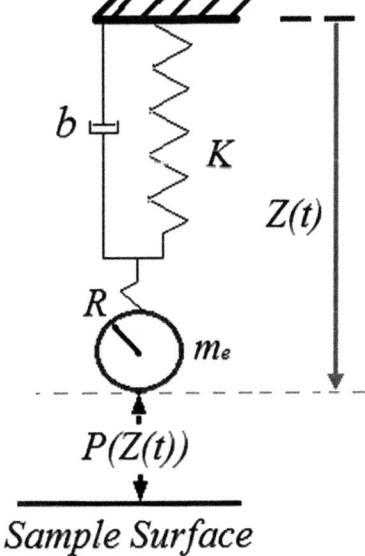

Fig. 3 Bond-graph model of
the AFM

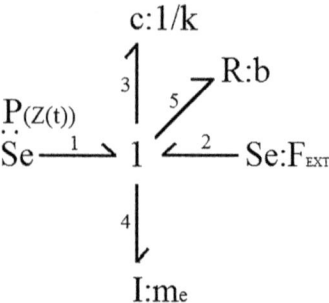

zero if power orientations are taken positive toward the junction. According to this, one can write the constitute law of the 1-junction as follows:

$$e1 + e2 - e3 - e4 - e5 = 0;$$
$$f1 = f2 = f3 = f4 = f5. \tag{2}$$

In a bond-graph, half arrows show the direction of the power. Causality shows the direction of the effort, and it is the basis for understanding the system's operation and modeling. Causality establishes the cause-and-effect relationship between the factors of the power of a bond. The information of the effort is represented by putting a small transverse stroke (causal stroke) at the end of the bond [9–13]. Figure 4 shows the causality for our system.

In our case, the external harmonic excitation and nonlinear tip-sample interaction force get in effort as starting information and determine the causality of the effort source. Elements C and I have integral causality. At a 1-junction, only one bond should bring the information about flow. Only one bond should be open-ended, and this constitutes the causality of resistive element R4.

The 20-sim program package allows for constructing and simulating models in the form of bond-graph models without using state-space equations.

The bond-graph model in 20-sim, given in Fig. 5, has further parameters: $m = 1$, $C = 1$, $R = 0.5$, $\eta = 0.2$, $w = 0.5$, $q = 1$. Dimensionless variables are used as described in [8].

Fig. 4 Causality bond-graph
model of the AFM

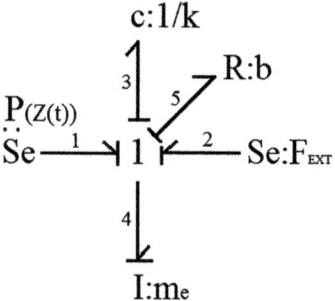

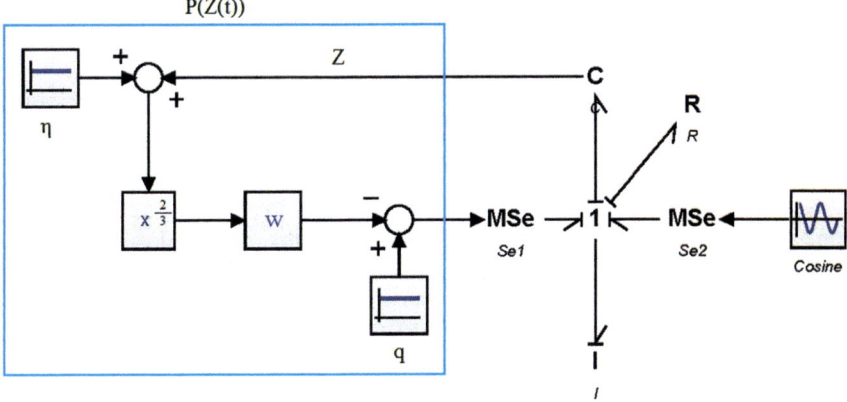

Fig. 5 Bond-graph model of the AFM in 20-sim

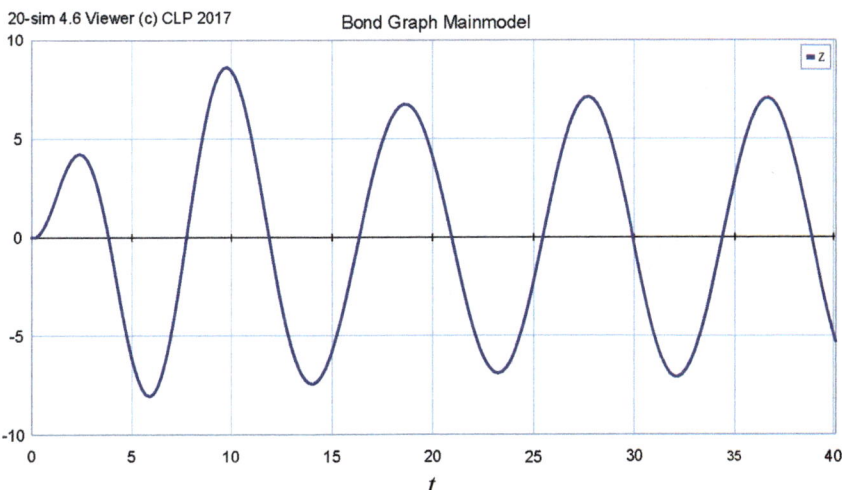

Fig. 6 The behavior of tapping mode AFM

The resulting motion of the AFM cantilever with external harmonic oscillation of amplitude 4 and excitation frequency $\Omega = 40$ is shown in Fig. 6.

Modeling of an AFM Cantilever Through MATLAB/SIMULINK

In order to check the validity of results obtained by 20-sim, the dynamic behavior of the system is studied through MATLAB/SIMULINK. The equation of motion of the AFM cantilever shown in Fig. 2 is

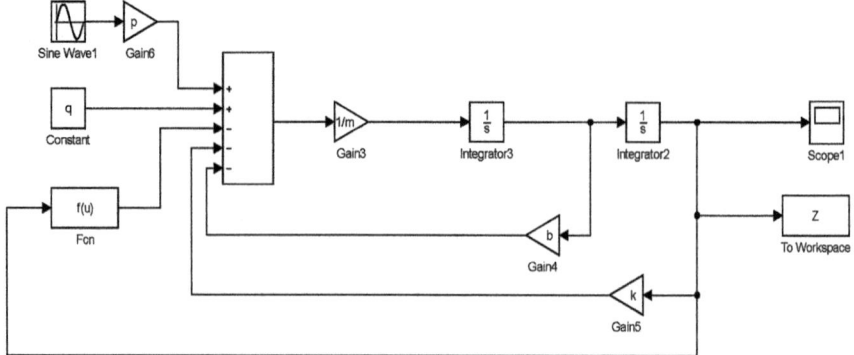

Fig. 7 Modeling of the AFM using MATLAB/SIMULINK

$$m_e\ddot{Z}(t) + b\dot{Z}(t) + KZ(t) = F_{EXT} - P(Z(t)). \tag{3}$$

We use the dimensionless equation of motion obtained in [8]:

$$\frac{d^2x}{d\tau^2} + b\frac{dx}{d\tau} + x = p\,\cos(\omega\tau) + q - w(\eta + x)^{3/2}. \tag{4}$$

We can represent (4) using MATLAB/SIMULINK, as shown in Fig. 7.

Setting the variables as described in the previous section, the result will be the same as that obtained by 20-sim through bond-graph modeling.

Conclusion

In this paper, bond-graph modeling of the Atomic force microscope was presented. The Atomic force microscope has been modeled as a lumped parameter system in its dynamic contact mode of operation. The Derjaguin–Muller–Toporov (DMT) force has been considered as the interaction of the cantilever tip with the sample surface. The response of the model is obtained through bond-graph by using the 20-sim program. Results were compared with results obtained through MATLAB/SIMULINK.

Acknowledgements The author would like to sincerely thank Professor V. V. Eliseev, who supported this research.

References

1. Binnig G, Quate CF, Geber C (1986) Atomic force microscope. Phys Rev Lett 56(9)
2. Materassi D, Basso M, Genesio R (2004) Frequency analysis of atomic force microscopes with repulsive-attractive interaction potentials. In: Proceedings of IEEE conference on decision and control, pp 3059–3061
3. Sebastian A, Salapaka MV, Chen DJ, Cleveland JP (2003) Harmonic analysis based modeling of tapping-mode AFM. In: Proceedings of American control conference, vol 1, pp 232–236
4. Wang L (1998) Analytical descriptions of the tapping-mode atomic force microscopy response. Appl Phys Lett 73(25):3781–3783
5. Gauthier M, Tsukada M (2000) Damping mechanism in dynamic force microscopy. Phys Rev Lett 85(25):5348–5351
6. Belikov S, Magonov S (2009) Classification of dynamic atomic force microscopy control modes based on asymptotic nonlinear mechanics. In: Proceedings of American control conference, pp 979–984
7. Bahrami MR, Ramezani A, Osgouie K (2010) Modeling and simulation of non-contact atomic force microscope. In: Proceedings of the ASME 2010 10th conference on engineering system design and analysis, ESDA 2010, vol 5, pp 565–569. https://doi.org/10.1115/esda2010-24394
8. Bahrami MR, Abeygunawardana AWB (2018) Modeling and simulation of dynamic contact atomic force microscope. In: 16th International symposium topical problems in the field of electrical and power engineering, Estonia, Kuressaare
9. Gawthrop PJ, Smith L (1996) Metamodelling: bond graphs and dynamic systems. Prentice Hall International, London
10. Thoma JU (1990) Simulation by bond graphs: introduction to a graphical method. Springer-Verlag, Berlin
11. Thoma J, Ould Bouamama B (2000) Modelling and simulation in thermal and chemical engineering: a bond graph approach. Springer, Berlin
12. Fakri A, Rocaries F, Carriere A (1997) A simple method for the conversion of bond graph models in representation by block diagrams. In: International conference on bond graph modeling and simulation (ICBGM'97), Phoenix, pp 15–19
13. Vidojkovic B, Antic D, Dankovic B (1997) Bondsim-simulink tools for bond graph modelling and simulation. In: Proceedings of 7th symposium of mathematics and its applications, Timisoara, 1997, pp 243–248

The Likelihood Description of Lubrication Layer Formation Structured at the Molecular Level

Elena V. Berezina, Alexej V. Volkov, Vladimir A. Godlevskiy, Alexander S. Parfenov and Anton G. Zheleznov

Abstract The mathematical model of boundary lubrication layer formation from a solution of a surface-active mesogene additive is elaborated. On that basis, the calculation of kinetic parameters of lubricant film formation is possible.

Keywords Tribology · Boundary lubrication layer · Mesogen · Surfactant Mathematical modelling

Introduction

In a case of absence of surface chemical processes, boundary lubricant layers are generated due to the influence of surface long-range forces and anisotropic and molecular interaction, both in the bulk of the lubricating material (LM) and during the process of the adsorptive binding of the lubricating material molecules with the solid surface. The presence of such layers provides the required durability of the solid surfaces and reduces the friction coefficient.

However, the friction boundary layer is not readily available for studying in situ due to its poor accessibility and dynamic changes of its thickness and inner structure. Only in recent times have control methods of the structural condition of

E. V. Berezina (✉) · A. V. Volkov · V. A. Godlevskiy · A. G. Zheleznov
Ivanovo State University, Ivanovo, Russia
e-mail: elena_berezina@mail.ru

A. V. Volkov
e-mail: volkz@ivanovo.ac.ru

V. A. Godlevskiy
e-mail: godl@yandex.ru

A. G. Zheleznov
e-mail: antonzhelezn@gmail.com

A. S. Parfenov
Ivanovo State Medical Academy, Ivanovo, Russia
e-mail: alsparf@gmail.com

© Springer International Publishing AG 2018
A. N. Evgrafov (ed.), *Advances in Mechanical Engineering*, Lecture Notes in Mechanical Engineering, https://doi.org/10.1007/978-3-319-72929-9_3

the lubricating layer during friction begun to be developed [1], as well as methods of molecular dynamics using computer molecular models of tribosystems with a lubricating layer on the nanoscopic level [2]. In general, the task is to describe the kinetics of the layer formation process, taking into account the molecular structure of the triboactive component of LM. In the long view, this will help to predict the tribological characteristics of the lubricating compositions, taking into account the mesomorphic parameters of the additives used.

Application of different types of amphiphile, such as the triboactive components of LM, allows for the carrying out of an approach for the description of the layer that has been formed in the theory of lyotropic liquid crystals [3], taking into account the «surface» -specific character of the real lubricating process.

Problem Statement

It is very difficult to perform an analytical description of the spatial characteristics of the ordered layer structure consisting of a large amount of molecules. However, there is a possibility of defining the layer from the aspect of the kinetics of its forming, as was done earlier by some of the authors of the current article concerning the description of the extreme lubricating process during the edge-cutting machining of metals [4, 5]. The characteristic time of ordering (final stage of layer creation) can serve as a characteristic of the process of forming the poly-molecular layer. In this paper, we propose an attempt to create such a description based on the timing data.

Mathematical Description

There is a certain set of one-type molecules located near the solid surface. It can be characterized by some value, which we will call *adsorption activity* A of the molecule. At the same time, we can imagine a multitude of solid surfaces, which should be characterized by some value; let us call it *adsorption passivity* S. Then, let us introduce the initial scale for the activity and passivity of the surface—zero. This means that zero activity is a point of reference.

Let us determine the adsorption activity of molecule A. Let $S_0 = 1$ be the pas-sivity of the surface taken as 1. Then,

$$A = \frac{P_{A_1}}{q_{A_1}}, \tag{1}$$

where P_{A1} is the probability of this molecule forming a bond with the standard surface and $q_{A_1} = 1 - P_{A_1}$ is the probability of it not forming a bond with the surface $S_0 = 1$.

Let us suppose that there is a surface characterized by the parameter $S_0 = 1$. Any molecule of the environment can collide with the surface. But at the same time, it can form (or not form) an adsorption bond with some probability. Let us determine the adsorption activity as the ratio of probability that the bond of this molecule will be formed on the surface S_0 to the probability that it won't be formed.

It follows from (1) that the higher the probability P_{A1}, the higher the activity of the molecule.

Let us determine the *adsorption passivity* of the surface S as the ratio of probability that this molecule will not form the bond with the surface to the probability of formation of this bond. Therefore, the higher the probability that the bond will be formed, the higher the passivity of the surface.

Let $A_0 = 1$ be the adsorption activity of a standard molecule. Then,

$$S = \frac{q_{S_1}}{P_{S_1}}, \tag{2}$$

where q_{S1} is the probability $A_0 = 1$ that the molecule will not bind and P_{S1} is the probability that it will.

Let us assume that if there are molecules with adsorption activities A_1 and A_2 and a surface with adsorption passivity S, then $\frac{A_1}{A_2} = \frac{P_{A_1 S}}{q_{A_1 S}} : \frac{P_{A_2 S}}{q_{A_2 S}}$ will not depend on S, where $P_{A_1 S}$ is the probability of the adsorption of a molecule with activity A_1 at the surface with passivity S. If we assume that $A_1 = A$, $A_2 = A_0 = 1$, then the probability that this molecule with activity A will bind with the surface with passivity S will appear as follows:

$$\frac{A}{1} = \frac{P_{AS}}{q_{AS}} : \frac{P_{1S}}{q_{1S}}. \tag{3}$$

Since $S = \frac{q_{1S}}{P_{1S}}$, then it can be set down that

$$\frac{A}{S} = \frac{P_{AS}}{q_{AS}} \rightarrow P_{AS} = q_{AS}\left(\frac{A}{S}\right). \tag{4}$$

Let us introduce a new value $\frac{A}{S} = \xi$, being the ratio of adsorption activity to adsorption passivity. Since $q_{AS} = 1 - P_{AS}$, it can be set down that

$$P_{AS} = \frac{\xi}{1+\xi}, \quad P_{AS} = P(A, S). \tag{5}$$

As can be seen from (5), we approach a one-parameter model. Only one parameter is important—the ratio of the molecule's activity to the surface passivity. It is reasonable to suggest that when the multi-layer adsorption films are generated, the first result will be that the films that structure and the molecules that are most likely to be bound will be generated. Consequently, if h is the thickness of an adsorption film, then $P(h) = P(\xi(h))$ is the probability of the film with thickness

h being generated. Here, $\xi(h)$ is the ratio of the molecule adsorption activity to the film's passivity, which also depends on the thickness of the generated film.

In order to continue with the adsorption film generation dynamics, let us formulate the following hypothesis: *the rate of film generation is proportional to the number of collisions with the surface of all molecules that the bulk of the lubricant film above this surface contains at that moment.* The surface is most likely to bind with those molecules with a higher adsorption probability for this surface. Consequently, the film generation dynamics can be set down as

$$\frac{dh}{dt} \approx P(h) \cdot n(t), \tag{6}$$

where $n(t)$ is the number of collisions of molecules with the surface unit per unit of time, i.e.,

$$\frac{dh}{dt} = \beta P(h) \cdot n(t), \tag{7}$$

where β is a coefficient of proportionality.

In order to give a comprehensive description of the generation dynamics of the film in which adsorption passivity is increasing while it is being filled, it is necessary to formulate a hypothesis of the behavior of the function $P(h)$. Let us assume that

$$-\frac{dP(h)}{dh} = \alpha P(h), \tag{8}$$

i.e., the rate of the decrease of the probability that the film with thickness h will be generated is proportionate to the probability of its generation, where α is the coefficient of proportionality. Then,

$$P(h) = P(0)e^{-\alpha h}, \tag{9}$$

where $P(0)$ is the first film generation probability.

From the last equation, we see that the passivity of the films eventually increases while the probability of their generation is decreasing. Considering (7) and (9), it can be set down that

$$\frac{dh}{dt} = \beta P(0)e^{-\alpha h}n(t), \tag{10}$$

$$\frac{dh}{e^{-\alpha h}} = \beta n(t)dt, \tag{11}$$

$$e^{\alpha h}dh = \beta n(t)dt, \tag{12}$$

$$\int e^{\alpha h} dh = \beta \int n(t) dt + C, \tag{13}$$

$$\frac{1}{\alpha} e^{\alpha h} = \beta \int n(t) dt + C, \tag{14}$$

At $t = 0$, $h = 0 \rightarrow C = \frac{1}{\alpha}$.

$$e^{\alpha h} = \alpha \beta \int n(t) dt + \alpha \cdot \left(\frac{1}{\alpha} \right), \tag{15}$$

$$\ln e^{\alpha h} = \ln \left(\alpha \beta \int n(t) dt + 1 \right), \tag{16}$$

$$\alpha h = \ln \left(\gamma \int n(t) dt + 1 \right), \tag{17}$$

where $\gamma = \alpha \cdot \beta$.

$$h = \frac{1}{\alpha} \ln \left(\gamma \int n(t) dt + 1 \right). \tag{18}$$

If $n(t) = n_0 = const$—i.e., n is kept constant, then

$$h = \frac{1}{\alpha} \ln(\gamma n_0 t + 1). \tag{19}$$

Above, we have defined the aim of the created model, which is to estimate the certain time of the generation of the structured film of hc in thickness in order to be able to include the typical adsorption time in the sum of times of the individual stages of the process.

If we suppose that during film generation, $P(h)$ will decrease e times, then the typical thickness of the film generated during a certain time τ will be equal to $h_0 = \frac{1}{\alpha}$. Consequently, the time of the film generation τ will be

$$\ln(\gamma n_0 \tau + 1) = 1, \tag{20}$$

$$\gamma n_0 \tau + 1 = e. \tag{21}$$

Therefore, we get a final formula allowing us to estimate the polymolecular film adsorption time:

$$\tau = \frac{e - 1}{\gamma n_0} \approx \frac{1.71}{\gamma n_0}. \tag{22}$$

The estimation of the collisions of molecules with the surface can be made using the formula

$$n_0 = \frac{1}{2} n \sqrt{\frac{iRT}{3\mu}}. \tag{23}$$

Parameter γ, can, possibly, depend on the degree of molecular orientation taking part in the formation of the layer. Since the model is probabilistic, then the same molecule can orient itself differently on the surface, and this should be taken into account in (23) as the structuring factor γ. Apparently, the meaning of this parameter can be obtained through the methods of computer molecular modeling [6]. From the formula for the calculation of friction boundary layer forming time, we see that the higher the value of γ, the less time is needed for formation of the layer.

An important moment in the current description is characteristic of the adsorption using the ration (4), which, however, requires interpretation in thermodynamic terms that we will try to do. The probabilistic character of the adsorption can be well seen in the works of Ya. I. Frenkel (see [7], for example). He has obtained an important expression for the dependence of molecule adsorption time τ (being the period of molecule vibration perpendicular to the adsorbent surface) on the molar heat capacity of adsorption Q_a and temperature T:

$$\tau = \tau_0 \exp(Q_a/RT), \tag{24}$$

where τ_0 is the time, close to the period of atom vibrations in the array of adsorbent, and has an order 10^{-13} c.

It's known that adsorption time ranges from 10^{-12} to 10^{-6} for physical adsorption to 10^{-2}–10^{-13} for chemical adsorption. Therefore, relying on the time intervals, we can introduce the notion of adsorption probability as the value proportional to the ratio of time that the molecule spends on the adsorbent surface τ to the characteristic time of observance for the process of adsorption or some time which limits the process of an adsorption layer forming (in our case, this can be the time of existence of a single interfacial capillary τ_k):

$$p \cong \frac{\tau}{\tau_k}. \tag{25}$$

Since it follows from the characteristic time of adsorption that this ratio can significantly exceed 1 (i.e., the ultimate value of probability), and in this case, the coefficient of proportionality will change in the broad range, then we will suggest that the ratio of these times is proportional to the ratio of adsorption probability p to the opposite event—probability q that there will not be adsorption on the surface q:

$$\frac{p}{q} = \alpha \frac{\tau}{\tau_k}, \tag{26}$$

where α is the dimensionless factor of proportionality.

Taking into account that $q = 1 - p$, the last formula can be written regarding p in the following way:

$$p = \frac{\frac{\alpha \tau}{\tau_k}}{1 + \frac{\alpha \tau}{\tau_k}} = \frac{1}{1 + \frac{\tau_k}{\alpha \tau}}. \tag{27}$$

Taking into account (24), this expression will be as follows:

$$p = \frac{1}{1 + \frac{\tau_k}{\alpha \tau_0} \exp(-Q_a/RT)}. \tag{28}$$

This equation shows interconnection of the adsorption probability with the molar heat capacity of adsorption and with the temperature.

Now we can define the notions of molecule activity A and surface passivity S [see (4)], the relation of which can be replaced by the ratio p/q:

$$\frac{A}{S} = \frac{p}{q}. \tag{29}$$

And taking into account (24) and (25), we find that their ratio equals

$$\frac{A}{S} = \alpha \frac{\tau_0}{\tau_k} \exp(Q_a/RT), \tag{30}$$

which testifies to the fact that the introduced characteristics of surface interaction of molecules make a certain physical sense.

Conclusion

Thus, we have reached a rather full description of the process of friction boundary layer formation from a liquid lubrication medium containing a triboactive adsorbing component. We have introduced structuring factor γ, which can be accessed in experimental physical-chemical studies. It should be noted that this model is connected with the properties of the layer (its solidity, shear characteristics, etc.,—this requires additional studies).

Another obvious and most significant limitation of the developed approach: the model does not take into account the possible chemical reactions of the LM components with the surface (for example, chemisorption of the oxygen dissolved in LM).

The nature of the connection between the found parameter γ and the structural parameter of mesomorphic structures Q_s mentioned above is also of interest. Of course, γ should depend on Q_s, at least due to the fact that under $Q_s = 0$, parameter γ makes no sense, and thus $\tau \to \infty$. It seems that the experimental determination of the parameter Q_s for the layer picture is also possible through the methods of molecular computer modeling.

References

1. Berezina EV, Volkov AV, Zheleznov AG, Fomichev DS (2015) On the research technique of mesogene lubrication layer optical properties. In: Evgrafov A (ed) Advances in mechanical engineering, lecture notes in mechanical engineering. Published by Springer International Publishing, Switzerland, pp 7–12. https://doi.org/10.1007/978-3-319-15684-2_2
2. Blinov OV, Godlevskiy VA (2016) Computing of lubrication layer molecular orientation state. Procedia Eng 150:584–589
3. Usol'tseva NV (1994) Lyotropic liquid crystals: chemical and supramolecular structure. Ivanovo, 220 p (In Russian)
4. Godlevskiy VA, Markov VV, Usoltseva NV (2017) Principle of compatibility of heterogeneous additives in trioactive metalworking fluids for edge cutting of metals. In: Evgrafov A (ed) Advances in mechanical engineering, lecture notes in mechanical engineering. Published by Springer International Publishing, Switzerland, pp 65–71. https://doi.org/10.1007/978-3-319-53363-6_8
5. Astakhov VP, Godlevskiy VA, Joksch S, Rave A, Evans R (2012) Metal working fluids for cutting and grinding: fundamentals and recent advances. Woodhead Publishing Ltd., Cambridge, 413 p
6. Kuznetsov SA Berezina EV, Godlevskiy VA, Bogomolov MV (2012) A software complex for molecular simulation of boundary lubrication layers. J Friction Wear 33(1):5–10
7. Fridrichsberg DA (1984) Colloid chemistry course. St.-Petersburg, "Chimija", 368 p. (In Russian)

Design of Library of Metaheuristic Algorithms for Solving the Problems of Discrete Optimization

Vladislav A. Chekanin and Alexander V. Chekanin

Abstract The paper contains a detailed description of major requirements which should be taken into account in the design of the library of metaheuristic algorithms of discrete optimization. The proposed requirements provide an opportunity to create a large number of different population-based algorithms by the designed library, including classical and modified variations of genetic algorithms, ant colony algorithms, bee algorithms, as well as many other optimization algorithms applicable for solving the problems of discrete optimization.

Keywords Metaheuristic algorithms · Evolutionary algorithms
Bionic algorithms · Algorithm library · Discrete optimization

Introduction

A large number of problems arising in the automation of design and management processes belong to the wide class of optimization problems, consisting in finding optimal solutions. Many optimization problems are the combinatorial optimization problems characterized by a variety of solutions of different quality [1]. Among the classic problems of combinatorial optimization, we can distinguish the traveling salesman problem, the knapsack problem, the resource allocation problem, the portfolio selection problem, the scheduling problem, the clique problem, the assignment problem, the graph coloring problem, and many other important problems [2–11]. These algorithmically complex problems have a wide range of practical applications and they are relevant to the automation of design and management processes in various industries and economies [12, 13].

V. A. Chekanin (✉) · A. V. Chekanin
Moscow State University of Technology «STANKIN», Moscow, Russia
e-mail: vladchekanin@rambler.ru

A. V. Chekanin
e-mail: avchekanin@rambler.ru

© Springer International Publishing AG 2018
A. N. Evgrafov (ed.), *Advances in Mechanical Engineering*, Lecture Notes
in Mechanical Engineering, https://doi.org/10.1007/978-3-319-72929-9_4

All of the combinatorial optimization problems belong to the class of NP-complete problems, and as a result, it is required that we use the resource-intensive optimization algorithms to obtain the optimal solutions to the problems, a task that proves to be inefficient in practice due to the large expenditure of time resources [1, 14]. One of the most prospective methods used to solve NP-complete problems of combinatorial optimization is the application of approximate metaheuristic algorithms of artificial intelligence, which do not guarantee obtaining an optimal solution, although they do allow us to obtain a set of suboptimal solutions of acceptable quality in a relatively short period of time. At present, the most efficient and interesting are the metaheuristic optimization algorithms based on bionic models borrowed from nature. Such algorithms include the genetic algorithms, the simulated annealing algorithm, the ant colony algorithm, the bee algorithm, the particle swarm algorithm, and the artificial immune systems [12, 13, 15–20]. The effectiveness of bionic algorithms is confirmed by a number of examples of their work in nature.

In this article, the most important principles for constructing a unified algorithmic base are presented in the form of a class library which will provide the possibility of realization of the well-known metaheuristic algorithms of artificial intelligence; this library will be able to create a variety of new hybrid and modified population algorithms based on bionic optimization ideas.

Requirements for the Library of Artificial Intelligence Algorithms

The library of artificial intelligence algorithms should serve as a basis for designing and researching new algorithms based on bionic optimization models. As a result of designing hybrid and modified algorithms of artificial intelligence, new optimization algorithms that have no analogs in nature can be created. Such algorithms will be obtained through the combination and adaptive adjustment of algorithms based on different bionic models.

The design of a unified library of artificial intelligence algorithms requires the defining of common parameters inherent to various metaheuristic optimization algorithms. Figure 1 shows the basic diagram demonstrating the process of solution of optimization problems using metaheuristic algorithms.

The application of the metaheuristic optimization algorithm requires selecting the following adjustable parameters.

1. Set of controllable parameters of the problem that most significantly affect the quality of the optimized solution (for example, in the packing problem, one such parameter is the sequence of selection of the objects to be placed).
2. Coding scheme for the controllable parameters, it is possible to use the following codes:

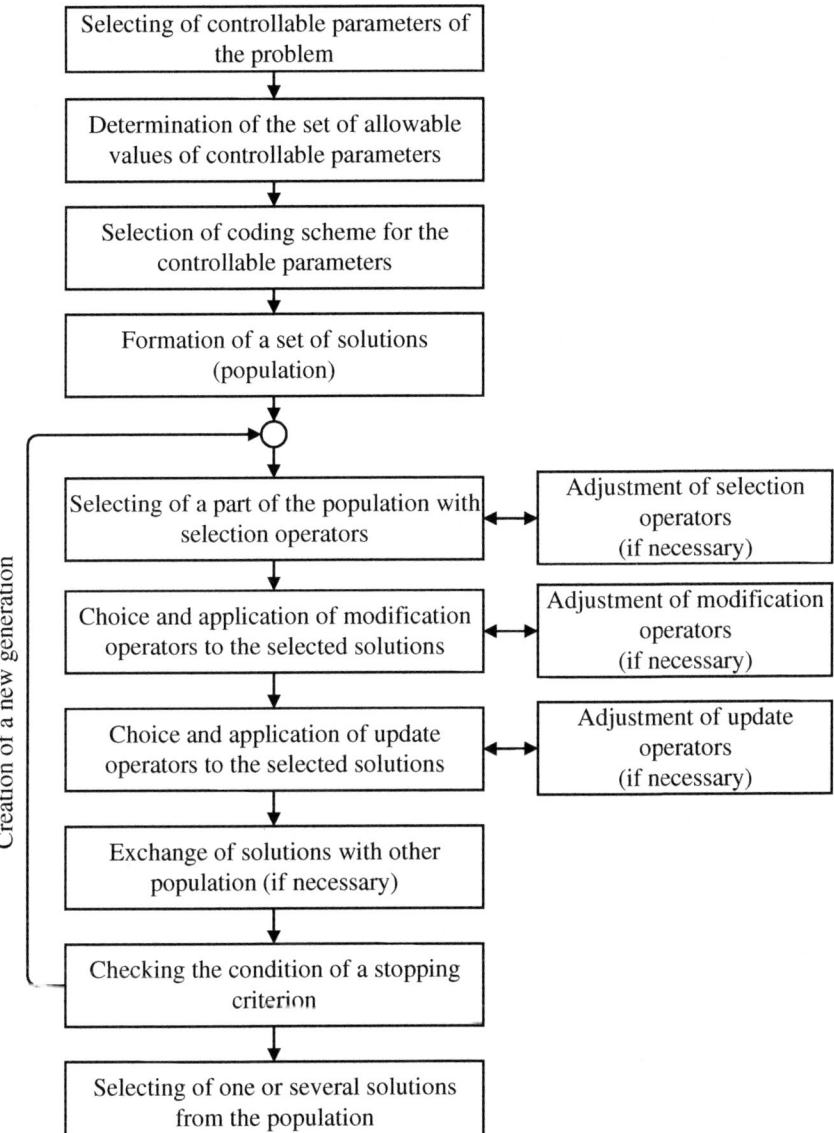

Fig. 1 The process of solving an optimization problem using a metaheuristic algorithm

- a binary code (the most effective in solving problems of continuous optimization);
- natural numbers (the most effective in solving problems of combinatorial optimization);
- other codes (Gray code, for example).

3. Optimization model which includes the following parameters:

    -   a set of unified simulated solutions, called a population;
    -   a set of unified simulated populations of solutions;
    -   rules for creating initial populations;
    -   a form of fitness function as a significant factor for estimation of the quality of solutions;
    -   rules for selecting subpopulations for modification of the solutions (the rules are given in the form of selection operators);
    -   operators for modification (differentiation) and updating, providing for the creation of new solutions and their inclusion in the population;
    -   rules for changing sets of the applied modification and update operators;
    -   a stopping criteria for the optimization process (among the most commonly used stopping criteria, the following can be distinguished: obtaining of the optimal solution, obtaining of a suboptimal solution of satisfactory quality, expiration of the time allotted for the solution of the problem).

Below are the main requirements defined as mandatory for the practical implementation of the universal library of artificial intelligence algorithms for solving optimization problems.

1. Requirements for the coded solutions:

    -   the ability to generate coded solutions of arbitrary length;
    -   the ability to change the lengths of the coded solutions (it is particularly necessary for realization of the multimethod genetic algorithm which optimizes the sequence of applied heuristics—the elementary methods of solving the problem [10, 15];
    -   the ability to specify a maximal lifetime for each solution.

2. Requirements for the population of solutions:

    -   the ability to join together solutions encoded by various methods;
    -   the ability to change the size of the population;
    -   the ability to set the maximum number of generations of the population;
    -   the ability to set the various rules for creating the initial population;
    -   the ability to set the various rules for selecting solutions for its modification;
    -   the ability to create the arbitrary subpopulations;
    -   the ability to create the individual sets of modification and update operators for each generation;
    -   the ability to estimate the average fitness of the population and analyze the dynamics of its change.

3. Requirements for the sets of populations:

    -   the ability to exchange the solutions between several populations;
    -   the ability to include a new population in the set;

- the ability to delete a population from the set;
- the ability to analyze the dynamics of the changes in the average fitness of the set.

4. Requirements for the selection operators:

- the ability to select a random set of solutions for its modification;
- the ability to apply various standard selection methods (roulette selection method, tournament selection method, rank selection method, panmixia, genotype and phenotypic inbreeding, outbreeding, selection based on a given scale);
- the ability to add new selection methods;
- the ability to join together several selection operators.

5. Requirements for the modification and update operators:

- the ability to apply the single-point, two-point and multi-point crossover operators to the coded solutions;
- the ability to uniformly cross the coded solutions by the given scheme;
- the ability to mutate various parts of the coded solutions;
- the ability to mutate according to a rule given in the form of a function;
- the ability to recalculate the fitness function of the selected solutions according to some rule;
- the ability to generate a new random solution;
- the ability to change the probability of any operator at any time.
- the ability to add new modification and update operators.

6. Requirements for the optimization algorithms (choice of strategies for finding the solutions):

- support of standard evolution models (the models based on evolution theories proposed by Charles Darwin, Jean-Baptiste Lamarck, Hugo de Vries, Karl Popper, Motoo Kimura);
- the ability to modify the standard evolution models in the solving process;
- the ability to adaptively adjust the strategies used to search for the optimal solution.

Design of the Library of Artificial Intelligence Algorithms

In the program version of the designed library of artificial intelligence algorithms, the proposed idea of setting rules for the distribution and exchange of solutions between populations in the process of optimization should be implemented. These rules can be specified in the form of schemes showing the direction of exchange of solutions and the percentage composition of the populations participating in the exchange. Figure 2 shows an example of a scheme for copying 10% of the best

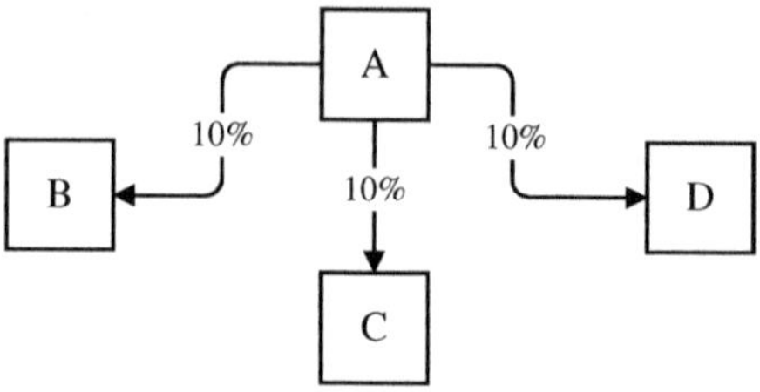

Fig. 2 Scheme of uniform
distribution of the best
solutions

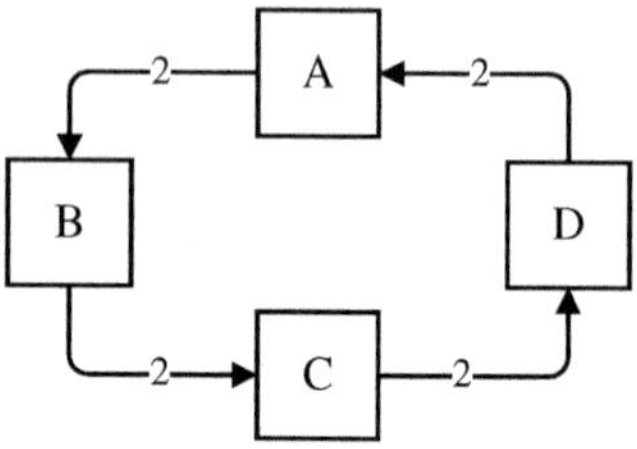

Fig. 3 Scheme of uniform
exchange of the best solutions
between populations

solutions from population A to populations B, C, and D. Figure 3 shows a scheme for copying the two best solutions from each population to the neighboring populations.

A designed universal library of artificial intelligence algorithms will support the following requirements for rules of exchange of solutions between populations:

- the ability to exchange the solutions;
- the ability to copy only the best solutions which are absent in the receiver's population;
- the ability to replace the worst solutions in the receiver's population;
- the ability to uniformly mix all of the solutions in the populations;
- the ability to copy all of the solutions to another population;
- the ability to change the exchange and distribution rules.

The proposed library of algorithms should provide the ability to create new combined optimization algorithms from several basic algorithms, thus forming ensembles of algorithms. When creating an ensemble of algorithms, it is necessary to select a set of jointly working algorithms, as well as to define rules for the exchange of solutions (migration rules) between the algorithms.

The simplest example of an ensemble of algorithms constructed on the basis of one optimization algorithm is a parallel genetic algorithm. Parallel genetic algorithms are based on the partitioning of the original population of solutions into several separate subpopulations, each of which is processed by a separate genetic algorithm independently of other subpopulations. In this algorithm, the exchange of solutions between subpopulations is possible only at the moment of transition to the next generation. Figure 4 gives an example of the combining of a set of genetic algorithms into a parallel genetic algorithm.

Fig. 4 A parallel genetic
algorithm as an example of an
ensemble of algorithms

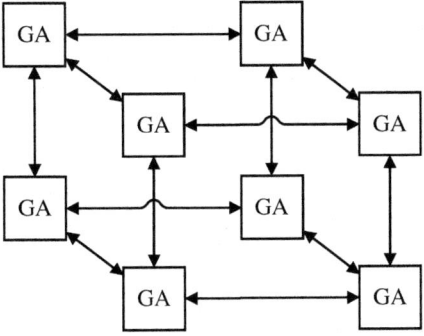

Ensembles of algorithms provide an opportunity to move beyond local optimums due to the cooperative usage of optimization algorithms that have different parameters of selection, modification and updating of their sets of populations.

The program implementation of the designed library is planned to be carried out using the object-oriented algorithmic language C++ and the standard template library (STL).

Conclusion

As a result of the research, the most significant requirements for the universal library of artificial intelligence algorithms based on the use of bionic optimization models were determined and classified. Further development of this research includes the software implementation of the described library of algorithms and its usage both for solving practical problems and for creating and analyzing hybrid and combined optimization algorithms.

Acknowledgement This work was carried out with the financial support of the Ministry of Education and Science of Russian Federation in the framework of the state task in the field of scientific activity of MSTU «STANKIN».

References

1. Johnson DS (2012) A brief history of NP-completeness, 1954–2012. Doc Math Extra Volume ISMP, pp 359–376
2. Bortfeldt A, Wascher G (2013) Constraints in container loading—a state-of-the-art review. EJOR 229(1):1–20
3. Chekanin AV, Chekanin VA (2014) Effective data structure for the multidimensional orthogonal bin packing problems. Adv Mater Res 962–965:2868–2871
4. Chekanin VA, Chekanin AV (2014) Improved data structure for the orthogonal packing problem. Adv Mater Res 945–949:3143–3146

5. Chekanin VA, Chekanin AV (2016) Implementation of packing methods for the orthogonal packing problems. J Theor Appl Inf Technol 88(3):421–430
6. Chekanin VA, Chekanin AV (2017) Deleting objects algorithm for the optimization of orthogonal packing problems. In: Advances in mechanical engineering. Springer International Publishing, pp 27–35
7. Chekanin VA, Chekanin AV (2015) An efficient model for the orthogonal packing problem. Adv Mech Eng 22:33–38
8. Chekanin VA, Chekanin AV (2016) New effective data structure for multidimensional optimization orthogonal packing problems. In: Advances in mechanical engineering. Springer International Publishing, pp 87–92
9. Martinez MAA, Clautiaux F, Dell'Amico M, Iori M (2013) Exact algorithms for the bin packing problem with fragile objects. Discret Optim 10(3):210–223
10. Valiahmetova YuI, Filippova AS (2007) Multi-method genetic algorithm for the decision of problems of orthogonal packing. Inf Technol (Informacionnye Tehnologii) 12:50–56 (in Russian)
11. Wascher G, Haubner H, Schumann H (2007) An improved typology of cutting and packing problems. EJOR 183(3):1109–1130
12. Karpenko AP (2012) Population algorithms for global continuous optimization. Review of new and little-known algorithms. In: Inf Technol (Informacionnye Tehnologii). vol 7. Appendix, 32 p (in Russian)
13. Shcherbina OA (2014) Metaheuristic algorithms for combinatorial optimization problems (Review). Tavricheskiy vestnik informatiki i matematiki 1:56–72 (in Russian)
14. Garey M, Johnson D (1979) Computers intractability: a guide to the theory of NP-completeness. W.H.Freeman, San Francisco, p 338
15. Chekanin VA, Chekanin AV (2014) Development of the multimethod genetic algorithm for the strip packing problem. Appl Mech Mater 598:377–381
16. Chernyak LS (2014) Intellekt roya dlya IT. Otkrytyye sistemy. SUBD 2:41–43 (in Russian)
17. Filippova AS (2006) Modeling of evolution algorithms for rectangular packing problems based on block structure technology. In: Inf Technol (Informacionnye Tehnologii). Appendix, 32 p (in Russian)
18. Gao YQ, Guan HB, Qi ZW, Hou Y, Liu L (2013) A multi-objective ant colony system algorithm for virtual machine placement in cloud computing. J Comput Syst Sci 79(8):1230–1242
19. Leung SCH, Zhang DF, Zhou CL, Wu T (2012) A hybrid simulated annealing metaheuristic algorithm for the two-dimensional knapsack packing problem. Comput Oper Res 39(1):64–73
20. Nseef SK, Abdullah S, Turky A, Kendall G (2016) An adaptive multi-population artificial bee colony algorithm for dynamic optimisation problems. Knowl Based Syst 104:14–23

Dynamics of the Manipulator Parallel-Serial Structure

Victor V. Dyashkin-Titov, Victor V. Zhoga, Ivan A. Nesmiyanov and Natalia S. Vorob'eva

Abstract The paper deals with a manipulator based on a tripod, a mechanism of parallel structure. As a gripping device, a manipulator of a sequential structure with three controllable degrees of freedom is used. Analytic expressions for the kinetic and potential energy of the manipulator are obtained. The technique of forming mathematical models of dynamics of a manipulator of a parallel-sequential structure with six degrees of freedom is developed. The task of positioning the manipulator's working body as it moves from the initial position to the specified final position is solved. Models of particular types of movement of the manipulator's working body make it possible to determine the program driving forces and moments necessary to implement the specified program movements of the working body. The study was carried out with the financial support of the Russian Foundation for Basic Research in the framework of scientific projects No. 16-38-00485 mole_a, 16-48-340395r_a.

Keywords Manipulator of parallel-sequential structure · Equations of dynamics
Capture device with three degrees of freedom

Introduction

In modern technological machines, mechanisms of parallel structure have spread. Manipulators of a parallel structure are characterized by high performance, reliability and accuracy of realization of program motions [1]. These qualities stipulate the use of such manipulators for machining parts of complex geometry [2, 3], instrument making, food packaging in the food industry [4], and in technological processes for the processing of agricultural products.

V. V. Dyashkin-Titov (✉) · I. A. Nesmiyanov · N. S. Vorob'eva
Volgograd State Agrarian University, Volgograd, Russia
e-mail: c_43.52.00@mail.ru

V. V. Zhoga
Volgograd State Technical University, Volgograd, Russia

© Springer International Publishing AG 2018
A. N. Evgrafov (ed.), *Advances in Mechanical Engineering*, Lecture Notes in Mechanical Engineering, https://doi.org/10.1007/978-3-319-72929-9_5

The most famous manufacturers of parallel manipulators are Fanuc, Yaskawa, Kawasaki, Omron (Japan), ABB (Sweden), Adept (USA), and Festo (Germany) [5].

However, these manipulators have a small service area and relatively limited manipulation of the gripping arm [6]. In the paper, a parallel-sequential structured manipulator is investigated, which allows us to increase the working area and the parameters of manipulation.

The Construction of the Manipulator

Figure 1 represents the kinematic scheme of the manipulator of a parallel-sequential structure. The manipulator consists of a spatial three-rod mechanism, in the form of a triangular pyramid with links of variable length l_1, l_2, l_3. Some ends of these links are connected by means of double-hinged joints to the base ABC, and the other ends are connected at the point M with a special hinge knot [7, 8]. The difference of the proposed manipulator scheme is the connection of linear actuators with the help of a spherical hinge ensuring the intersection of the geometric axes of the actuators at the point. This connection eliminates the appearance of bending moments on the links of the manipulator from external loads and provides high rigidity of the structure.

On the hinged assembly is mounted a controlled gripping device consisting of three links connected in series by kinematic pairs of the V class of links [9]. Electric cylinders are used as actuators.

Figure 2 shows the kinematic diagram of the capture device of the sequential structure, attached at point M to the manipulator—tripod.

The proposed manipulator scheme allows us to increase the parameters of manipulation, to provide the required orientation of the gripping device's working element at each point of the service area [10].

One of the basic requirements determining the operability of manipulators is to provide the approach of the manipulator's working body to the required points of the service object with a given orientation. Important tasks in the development of such manipulators are kinematic and dynamic analysis, the development of methods for the synthesis of displacements of actuating mechanisms, and determination of the values of the program control forces of the executive links when the working member is moved [11, 12].

The movement of the manipulator is considered with respect to the absolute coordinate system associated with the fixed base. With each link of the manipulator capture, mobile coordinate systems are connected. The axis of each link is directed along the axis of relative rotation of this link [13, 14].

As generalized coordinates, parameters are accepted that describe the configuration of the parallel structure manipulator (Fig. 1) and coordinates that describe the state of the capture device with three degrees of freedom (Fig. 2).

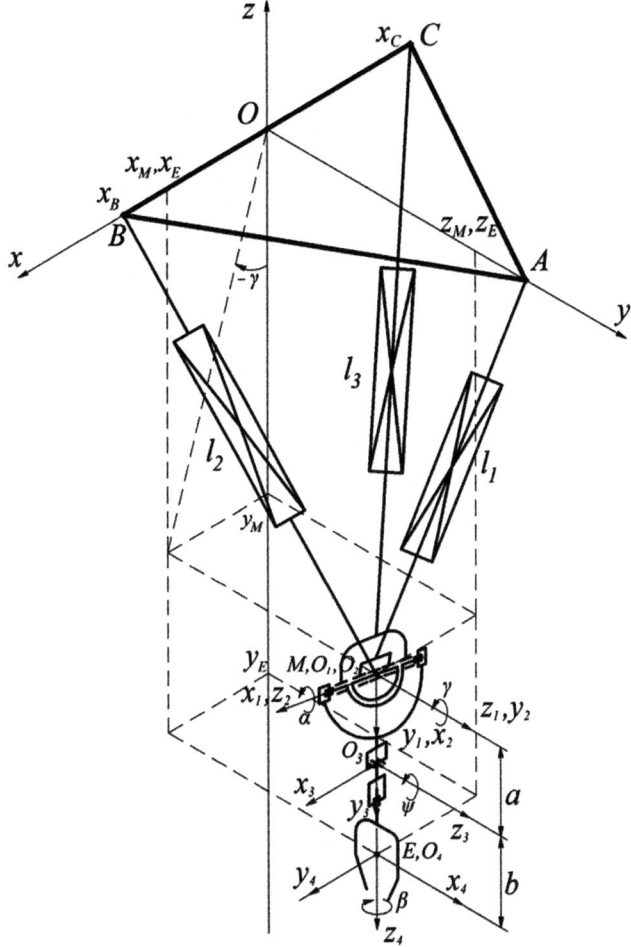

Fig. 1 Kinematic scheme of the manipulator of parallel-sequential structure

The manipulator design provides holonomic links between coordinates $q_1 = \gamma(t)$, $q_2 = x_M(t)$, $q_3 = y_M(t)$, $q_4 = z_M(t)$

$$f_1(q_s) = \gamma - \text{arctg}\,\frac{x_M}{z_M} = 0 \tag{1}$$

and link lengths $l_k, (k = 1 \div 3)$

$$
\begin{aligned}
f_2(q_s) &= \sqrt{x_M^2 + (y_M - y_A)^2 + z_M^2} - l_1 = 0,\\
f_3(q_s) &= \sqrt{(x_M - OB)^2 + y_M^2 + z_M^2} - l_2 = 0,\\
f_4(q_s) &= \sqrt{(x_M + OB)^2 + y_M^2 + z_M^2} - l_3 = 0.
\end{aligned}
\tag{2}
$$

Fig. 2 Kinematic scheme for
the capture of a tripod
manipulator with three
degrees of mobility

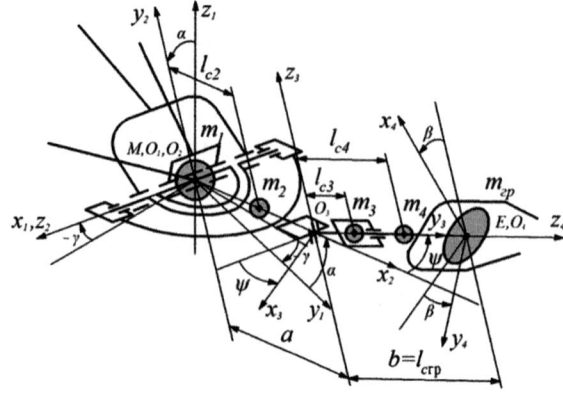

Thus, the number of independent generalized coordinates is 6.

The kinetic energy of a mechanical system subject to stationary connections is represented by a quadratic form of generalized velocities [15]

$$T = \frac{1}{2} \sum_{s=1}^{7} \sum_{k=1}^{7} A_{sk} \dot{q}_s \dot{q}_k, \quad A_{sk} = A_{ks}, \tag{3}$$

where $A_{sk} = A_{ks}$ are the coefficients depending on the generalized coordinates.

The values of the coefficients in expression (3), when the working member moves along an arbitrary trajectory, are equal:

$$A_{11} = I_y, \quad A_{22} = (G + m_{gr}), \quad A_{33} = (G + m_{gr}),$$
$$A_{44} = (G + m_{gr}), \quad A_{55} = I_\alpha, \quad A_{66} = I_\psi, \quad A_{77} = I_\beta,$$

$$A_{12} = A_{21} = (m_3 l_{c3} + m_4 l_{c4} + m_{gr} l_{cgr}) \sin \psi \sin \gamma$$
$$- \left[\begin{array}{l} m_2 l_{c2} + m_3(a + l_{c3} \cos \psi) + m_4(a + l_{c4} \cos \psi) \\ + m_{gr}(a + l_{cgr} \cos \psi) \end{array} \right] (\cos \alpha \cos \gamma),$$

$$A_{13} = A_{31} = -(m_3 l_{c3} + m_4 l_{c4} + m_{gr} l_{cgr}) \sin \psi \cos \gamma$$
$$+ \left[\begin{array}{l} m_2 l_{c2} + m_3(a + l_{c3} \cos \psi) + m_4(a + l_{c4} \cos \psi) \\ + m_{gr}(a + l_{cgr} \cos \psi) \end{array} \right] (\cos \alpha \sin \gamma),$$

$$A_{15} = A_{51} = m_3 l_{c3}(a + l_{c3} \cos \psi) \sin \alpha \sin \psi + (J_{3x} - J_{3y}) \sin \psi \cos \psi \sin \alpha$$
$$+ m_4 l_{c4}(a + l_{c4} \cos \psi) \sin \alpha \sin \psi + (J_{4y} - J_{4x}) \sin \beta \cos \psi \cos \beta \cos \alpha$$
$$- J_{4z} \sin \psi \sin \alpha + (J_{gry} - J_{grx}) \sin \beta \cos \psi \cos \beta \cos \alpha$$
$$+ m_{gr} l_{cgr}(a + l_{cgr} \cos \psi) \sin \alpha \sin \psi,$$

$$A_{25} = A_{52} = \begin{bmatrix} m_2 l_{c2} + m_3 (a + l_{c3} \cos \psi) + m_4 (a + l_{c4} \cos \psi) \\ + m_{gr} (a + l_{cgr} \cos \psi) \end{bmatrix} (\sin \alpha \sin \gamma),$$

$$A_{16} = A_{61} = \begin{bmatrix} m_3 l_{c3} (l_{c3} + a \cos \psi) + m_4 l_{c4} (l_{c4} + a \cos \psi) \\ + m_{gr} l_{cgr} (l_{cgr} + a \cos \psi) + J_{3z} + J_{4x} \cos^2 \beta \\ + J_{4y} \sin^2 b + J_{grx} \cos^2 \beta + J_{gry} \sin^2 \beta \end{bmatrix} \cos \alpha,$$

$$A_{26} = A_{62} = (m_3 l_{c3} + m_4 l_{c4} + m_{gr} l_{cgr})(\sin \psi \cos \alpha \sin \gamma - \cos \psi \cos \gamma),$$

$$A_{17} = A_{71} = I_\beta \sin \alpha, \; A_{57} = A_{75} = -I_\beta \sin \psi,$$

$$A_{53} = A_{35} = -\begin{bmatrix} m_2 l_{c2} + m_3 (a + l_{c3} \cos \psi) + m_4 (a + l_{c4} \cos \psi) \\ + m_{gr} (a + l_{cgr} \cos \psi) \end{bmatrix} (\sin \alpha \cos \gamma),$$

$$A_{36} = A_{63} = (m_3 l_{c3} + m_4 l_{c4} + m_{gr} l_{cgr})(- \sin \psi \cos \alpha \cos \gamma - \cos \psi \sin \gamma),$$

$$A_{45} = A_{54} = \begin{bmatrix} m_2 l_{c2} + m_3 (a + l_{c3} \cos \psi) + m_4 (a + l_{c4} \cos \psi) \\ + m_{gr} (a + l_{cgr} \cos \psi) \end{bmatrix} \cos \alpha,$$

$$A_{46} = A_{64} = - (m_3 l_{c3} + m_4 l_{c4} + m_{gr} l_{cgr}) \sin \psi \sin \alpha,$$

$$A_{56} = A_{65} = (I_{4y} - I_{4x}) \sin \beta \cos \beta \cos \psi + (I_{gry} - I_{grx}) \sin \beta \cos \beta \cos \psi, \quad (4)$$

where m_1 is the reduced mass of the manipulator to point M; m_2, m_3, m_4 are the masses of the links of ME; m_{gr} is the weight of the load; $G = m_1 + m_2 + m_3 + m_4$ is the weight manipulator; $l_{c2}, l_{c3}, l_{c4}, l_{cgr}$ are the distances to the centers of mass of the links of capture (Fig. 2) and to the center of mass of the carried load; $I_\gamma, I_\alpha, I_\beta$ are the moments of inertia of the manipulator reduced to the axes of relative rotation; $I_{ix}, I_{iy}, I_{iz}, \; i = 3, 4$ are the moments of inertia of the links 3 and 4 of the ME grip and the load carried relative to their own coordinate axes, respectively; $I_{grx}, I_{gry}, I_{grz}$ are the moments of inertia of the carried load relative to the coordinate axes x_4, y_4, z_4, respectively; and a is the distance between the mass of the point and the axis of rotation of the link 3.

All of the other coefficients in expression (3) are equal to zero. The given moments of inertia are calculated by the formulas

$$I_\gamma = \left(m_3 l_{c3}^2 + m_4 l_{c4}^2 + m_{gr} l_{cgr}^2 \right) \sin^2 \psi + I_{1z}$$
$$+ \left(I_{2x} + I_{3x} \sin^2 \psi + I_{3y} \cos^2 \psi + I_{4z} + I_{grz} \right) \sin^2 \alpha$$
$$+ \begin{bmatrix} I_{2y} + I_{3z} + m_2 l_{c2}^2 + m_3 (a + l_{c3} \cos \psi)^2 + m_4 (a + l_{c4} \cos \psi)^2 \\ + m_{gr} (a + l_{cgr} \cos \psi)^2 + (I_{4x} + I_{grx}) \cos^2 \beta + (I_{4y} + I_{gry}) \sin^2 \beta \end{bmatrix} \cos^2 \alpha,$$

$$(5)$$

$$I_\alpha = I_{2z} + m_2 l_{c2}^2 + m_3 (a + l_{c3} \cos \psi)^2 + m_4 (a + l_{c4} \cos \psi)^2$$
$$+ m_{gr}(a + l_{gr} \cos \psi)^2 + \left(I_{3y} + I_{4z} + I_{grz}\right) \sin^2 \psi$$
$$+ \left[I_{3x} + \left(I_{4x} + I_{grx}\right) \sin^2 \beta + \left(I_{4y} + I_{gry}\right) \cos^2 \beta\right] \cos^2 \psi,$$

$$I_\psi = I_{3z} + m_3 l_{3c}^2 + m_4 l_{4c}^2 + m_{gr} l_{cgr}^2 + \left(I_{4x} + I_{grx}\right) \cos^2 \beta + \left(I_{4y} + I_{gry}\right) \sin^2 \beta,$$

$$I_\beta = I_{4z} + I_{grz} = const,$$

where $I_{ix}, I_{iy}, I_{iz}, \quad i = 1, 2$ are the moments of inertia of the five-moving spherical hinge unit and the link 2 of the *ME* grip with respect to their own coordinate axes, respectively.

The potential energy of the manipulator has the form

$$\prod (q_k) = \left(m_3 l_{c3} + m_4 l_{c4} + m_{gr} l_{cgr}\right) g \sin \psi \sin \gamma$$
$$+ \left(G + m_{gr}\right) g z_M - \left[m_2 l_{c2} + \left(m_3 + m_4 + m_{gr}\right) a \right. \tag{6}$$
$$+ \left. \left(m_3 l_{c3} + m_4 l_{c4} + m_{gr} l_{cgr}\right) \cos \psi\right] g \cos \alpha \cos \gamma.$$

The Lagrange equations with indeterminate factors are explicitly written in the form [15]

$$\sum_{k=1}^{7} A_{sk} \ddot{q}_k + \sum_{k=1}^{7} \sum_{m=1}^{7} [k, m, s] \cdot \dot{q}_k \dot{q}_m = Q_s + \sum_{i=1}^{4} \lambda_i \frac{\partial f_i(q_s)}{\partial q_s} - \frac{\partial \prod}{\partial q_s}, \tag{7}$$

where Q_s is the generalized force of the control forces corresponding to the coordinate S; $-\frac{\partial \prod}{\partial q_s}$ is the generalized force of potential active forces, corresponding to the coordinate; λ_i are the Lagrange multipliers; and $[k, m, s]$ are the Christoffel symbols of the first kind

$$[k, m, s] = \frac{1}{2} \left(\frac{\partial A_{ks}}{\partial q_m} + \frac{\partial A_{ms}}{\partial q_k} - \frac{\partial A_{km}}{\partial q_s}\right). \tag{8}$$

To determine the partial derivatives entering into the expressions (8), a program of calculation using the apparatus of symbolic mathematics Mathcad was developed.

Synthesis of Program Displacements

The derivation of the analytic equations of motion of a manipulator with the help of the Lagrange formalism (7) for a system with six degrees of freedom is one of great complexity. However, most technological processes can be movements of the gripping device along special trajectories. Such trajectories satisfy fairly simple equations of dynamics, allowing for real-time control of the drives. The task of moving the manipulator is to determine the laws for changing the lengths of the executive links $l_k(t), (k = 1 \div 3)$ of the manipulator and the angles of rotation of the links of the gripping device $\alpha(t), \psi(t)$.

We assume that at the initial moment of time $(t_0 = 0)$, the manipulator configuration is known, determined by the values of the lengths of the executive links $l_k(0), (k = 1 \div 3)$ and the values of the generalized coordinates $q_5 = \alpha(0)$, $q_6 = \psi(0)$.

The spatial position of the working element at the current time t is completely determined by the coordinates $x_E(t), y_E(t), z_E(t)$ in the fixed coordinate system $Oxyz$ and by the direction cosines a_{pq} (Fig. 3) [16].

The position of the working member is determined by the product of the transition matrices $\mathbf{M_{i-1,i}}$, describing the position of the i link relative to $(i - 1)$ [17, 18]

$$\mathbf{M_{04}} = \mathbf{M_{01}} \times \mathbf{M_{12}} \times \mathbf{M_{23}} \times \mathbf{M_{34}}. \tag{9}$$

The relationship between the coordinates of points $\mathbf{x_i}$ in i and $\mathbf{x_{i-1,i}}$ in $(i - 1)$ coordinate systems is performed by sequential execution of operations

$$\mathbf{x_{i-1}} = \mathbf{M_{i-1,i}} \, \mathbf{x_i},$$

using transition matrices $\mathbf{M_{i-1,i}}$.

With a successive transition along the chain of hinged-connected links from the coordinate system $O_4x_4y_4z_4$ to zero $Oxyz$, applying expression (9), we obtain

$$\mathbf{M_{04}} = \begin{bmatrix} a_{11} & a_{12} & a_{13} \\ a_{21} & a_{22} & a_{23} \\ a_{31} & a_{32} & a_{33} \\ 0 & 0 & 0 \end{bmatrix} \begin{bmatrix} x_M + b_{13} - a\cos\alpha\sin\gamma \\ y_M + ba_{23} + a\cos\alpha\cos\gamma \\ z_M + ba_{33} + a\sin\alpha \\ 1 \end{bmatrix}, \tag{10}$$

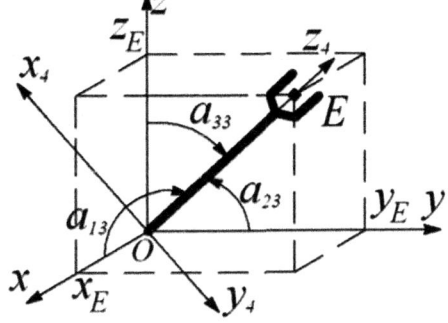

Fig. 3 Orientation of the working member of the manipulator grip device in the space

where a_{pq} are the direction cosines defining the orientation of the working member of the gripping device in space:

$$a_{13} = -\cos\gamma\sin\psi - \sin\gamma\cos\psi\cos\alpha;$$
$$a_{23} = \cos\gamma\cos\psi\cos\alpha - \sin\gamma\sin\psi; \qquad (11)$$
$$a_{33} = \cos\psi\sin\alpha$$

Equations (10) and (11) allow for determining the coordinates of the attachment point $M(x_M, y_M, z_M)$ of the gripping device to the tripod and the constructive angle γ. Then, the lengths (2) of the manipulator's executive links $l_k(t)$, $(k = 1 \div 3)$ are calculated, as well as the generalized coordinates $\alpha(t)$, $\psi(t)$ [19]. Thus, at the current time, the required manipulator configuration is fully known. The angle $\beta(t)$ does not affect the orientation of the working element, and the law of its change depends on the type of technological operation performed.

For example, when a manipulator is used as part of an automatic product packaging line, it is necessary to move the goods from one conveyor to another. Such an operation can be carried out by moving the center of mass of the cargo along an arc of a circle according to a given law of changing the arc coordinate $S(t)$. The equation of a circle of radius R, the plane of which is parallel to the fixed plane xOz (Fig. 1) $y_E = y_M = const$ with the center at the point $D(0, y_D, z_D)$, has the form

$$x_E^2 + (z_E - z_D)^2 = R^2. \qquad (12)$$

Points with coordinates $x_E(0) = x_{E0} = -x_{Ek}$, $z_E(0) = z_{E0} = z_{Ek}$ limit the arc of the circle to the same $S(\tau)$. The central angle corresponding to this arc is equal to $2\vartheta_0$. The laws of changing the Cartesian coordinates of the point $E[x_E(t), z_E(t)]$ at the current time are obtained in the form

$$x_E(t) = R\sin(\vartheta_0 - \vartheta), \quad z_E(t) = z_D + R\cos(\vartheta_0 - \vartheta), \quad \vartheta = \frac{S(t)}{R}. \qquad (13)$$

We assume that the working member moves translationally parallel to the axis xO, i.e., the direction cosines (Fig. 3) are constant and equal $a_{13} = -1$, $a_{23} = 0$, $a_{33} = 0$. Then, it follows from (11) that $\psi = \frac{\pi}{2} - \gamma$, $\dot{\psi} = -\dot{\gamma}$, $\alpha = 0$. From Eq. (10), we find

$$x_M = x_E + b + a\sin\gamma, \quad z_M = z_E, \quad tg\gamma = \frac{(x_E + b)}{y_E - y_A}. \qquad (14)$$

The expression (3) of the kinetic energy of the manipulator in this case has the form

$$T = \frac{1}{2}(G + m_{gr})(\dot{x}_M^2 + \dot{z}_M^2) + \frac{1}{2}I_0\dot{\psi}^2 + I_{x\psi}\dot{x}_M\dot{\psi}\sin\psi, \qquad (15)$$

where the following notation is introduced:

$$I_o = I_{1z} + I_{2y} = const; \quad I_{x\psi} = m_2 l_{c2} + a(m_3 + m_4 + m_{gr}) = const.$$

Differential equations for the motion of the manipulator (7) when moving a point $M(x_M, y_M, z_M)$ in the space of the working zone, taking into account the coupling equations (1) and (2) and equality (6), take the form

$$(G + m_{gr})\ddot{x}_M + I_{x\psi}\ddot{\psi}\sin\psi + I_{x\psi}\dot{\psi}^2\cos\psi$$
$$= \lambda_1 \frac{x_M}{l_1} + \lambda_2 \frac{(x_M - OB)}{l_2} + \lambda_3 \frac{(x_M + OB)}{l_3} - \lambda_4 \frac{z_M}{x_M^2 + z_M^2}, \quad (16)$$

$$0 = \lambda_1 \frac{y_M - y_A}{l_1} + \lambda_2 \frac{y_M}{l_2} + \lambda_3 \frac{y_M}{l_3}, \quad (17)$$

$$(G + m_{gr})\ddot{z}_M = \lambda_1 \frac{z_M}{l_1} + \lambda_2 \frac{z_M}{l_2} + \lambda_3 \frac{z_M}{l_3} + \lambda_4 \frac{x_M}{x_M^2 + z_M^2} - (G + m_{gr})g, \quad (18)$$

$$I_0 \cdot \ddot{\psi} + I_{x\psi}\ddot{x}_M \sin\psi = -\lambda_4, \quad (19)$$

where $\lambda_k = F_k, k = 1, 2, 3$ are the control forces in the linear links of the manipulator; and $\lambda_4 = M_\psi$ is the driving moment of the second link of the gripping device.

The law of changing the arc coordinate of the motion of a point E along the circle $S(t)$ is given in the form

$$S(t) = S(\tau)\left(10\frac{t^3}{\tau^3} - 15\frac{t^4}{\tau^4} + 6\frac{t^5}{\tau^5}\right). \quad (20)$$

This law ensures the movement of a point with zero values of speed and acceleration at the initial and final moments of the movement time.

From Eq. (14), we determine the laws of variation of the coordinates $x_M(t), z_M(t)$ and the angle $\gamma(t)$ and $\psi(t)$. Then, Eq. (2) allow us to determine the laws of variation in the lengths of the executive links $l_k(t), (k = 1 \div 3)$. From Eqs. (16)–(19), we find the program forces in linear actuators $F_k(t), (k = 1 \div 3)$, the uncertain Lagrange multiplier λ_4 and the M_ψ driving moment. Figure 4 shows the laws of the change in control forces in the linear executive links of the manipulator—tripod.

The following values of the manipulator parameters are accepted for calculations: mass—$m_1 = 30$ kg, $m_2 = 0.2$ kg, $m_3 = 0.3$ kg, $m_4 = 0.5$ kg, $m_{gr} = 2$ kg. Coordinates of the starting point—E_0–$x_{E0} = 213.6$ mm, $y_{E0} = 106.3$ mm, $z_{E0} = -886.1$ mm, the coordinates of the point of the end point—E_k–$x_{Ek} = -213.6$ mm, $y_{Ek} = 106.3$ mm, $z_{Ek} = -886.1$ mm, the angles $\alpha(0) = 0$, $\psi(0) = \frac{\pi}{2}$, $\beta(0) = 0$.

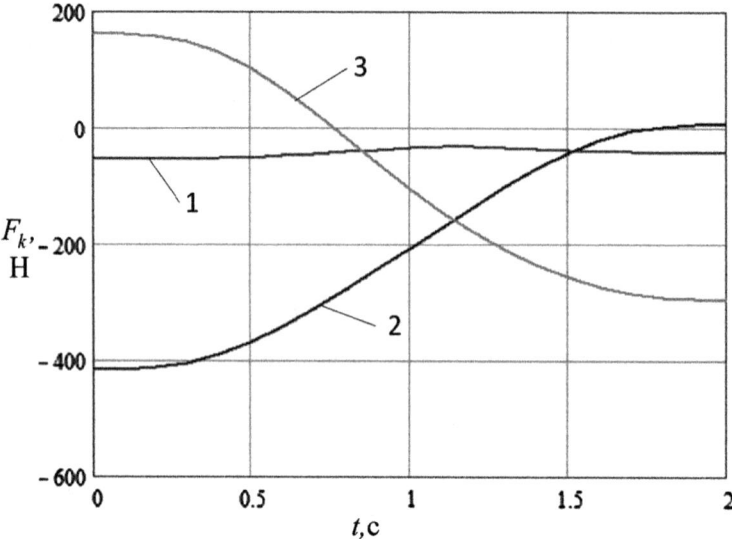

Fig. 4 Laws for changing the program effort in the executive units of the tripod manipulator

Conclusion

A technique for forming a knowledge base of models and manipulator controls with six degrees of freedom is developed. The knowledge base is filled with trajectories and equations of dynamics, based on the analysis of technological processes.

The knowledge base is the basis on which the intelligent control system of manipulator movement with associative memory is built. The resulting equations of the dynamics of the manipulator of a parallel-sequential structure allow for solving two basic problems—the determination of the driving forces and the moments necessary for the realization of the given program motions of the working member, as well as the efforts in kinematic pairs.

References

1. Glazunov VA, Koliskor AW, Krainev AF (1991) Spatial mechanisms of parallel structure. Nauka, Moscow, p 95 (in Russian)
2. Afonin VL, Podzorov PV, Slepcov VV (2006) Processing equipment on the basis of mechanisms of parallel kinematics (Obrabatyvajushhee oborudovanie na osnove mehanizmov parallel'noj kinematiki). Mashinostroenie, Moscow, p 448 (in Russian)
3. Bushuev VV, Hol'shev IG (2001) The mechanisms of parallel structure in mechanical engineering (Mehanizmy parallel'noj struktury v mashinostroenii). STIN. No. 1, pp 3–8 (in Russian)

4. The handling robot FlexPicker IRB 360 company AAB. www.aab.com/robotics [Electronic resource]. http://www.roboticturnkeysolutions.com/robots/abb/datasheet/IRB_360.pdf
5. Rybak LA, Grinenko GP (2013) Innovative processing equipment on the basis of parallel structures: prospects and directions of commercialization (Innovacionnoe obrabatyvajushhee oborudovanie na baze parallel'nyh struktur: perspektivy i napravlenija kommercializacii). Naukoemkietehnologii v mashinostroenii 7(25):32–39 (in Russian)
6. Kobrinskij AA, Kobrinskij AE (1985) Manipulation systems of robots (Manipuljacionnye sistemy robotov). Nauka, Moscow, p 343 (in Russian)
7. Gerasun VM, Pyndak VI, Nesmiyanov IA, Dyashkin-Titov VV, Pavlovskij VE (2012) Manipulators for mobile robots. The concept and design principles (Manipuljatory dlja mobil'nyh robotov. Koncepcii i principy proektirovanija), Preprinty IPM im. MV Keldysha, No. 44, p 24 (in Russian)
8. Zhoga V, Gavrilov A, Gerasun V, Nesmianov I, Pavlovsky V, Skakunov V, Bogatyrev V, Golubev D, Dyashkin-Titov V, Vorobieva N (2014) Walking mobile robot with manipulator-tripod. In: Proceedings of Romansy 2014 XX CISM-IFToMM symposium on theory and practice of robots and manipulators. Mechanisms and machine science, vol 22. Springer International Publishing, Switzerland, pp 463–471
9. Zhoga VV, Dyashkin-Titov VV, Dyashkin AV, Vorob'eva NS, Nesmiyanov IA, Ivanov AG (2017) RF Patent 2616493 IPC7 B66C 23/44. Byull. Izobret, 2017, no. 11 (in Russian)
10. Gerasun VM, Zhoga VV, Nesmiyanov IA, Vorobjeva NS, Dyashkin-Titov VV (2013) Mobile manipulator—tripod maintenance zone definition. Eng Mod Educ 3:2–8 (in Russian)
11. Nesmiyanov I, Zhoga V, Skakunov V, Terekhov S, Vorob'eva N, Dyashkin-Titov V Al-hadsha FAH (2015) Synthesis of control algorithm and computer simulation of robotic manipulator-tripod. Communications in computer and information science. Springer International Publishing, Switzerland, CIT&DS, CCIS 535, pp 392–404
12. Nesmiyanov IA, Zhoga VV, Skakunov VN, Vorob'eva NS, Dyashkin-Titov VN, Bocharnikov VS (2017) On the unstable operating modes of manipulator electric drives. J Mach Manuf Reliab 46(3):232–239
13. Vorobeva NS, Dyashkin-Titov VV, Zhoga VV, Nesmiyanov IA (2017) The dynamics of the manipulator parallel-serial structure based tripod. Mashinostr. Inzh. Obrazovan. No. 3, pp 32–41 (in Russian)
14. Dyashkin-Titov VV, Zhoga VV, Nesmiyanov IA, Vorob'eva NS (2017) Dynamics of the manipulator parallel-serial structure (Dinamika manipuljatora parallel'no-posledovatel'noj struktury). Sovremennoe mashinostroenie: Nauka i obrazovanie: Materialy 6-j Mezhdunar. nauch.-prakt. konferencii/ Pod red. AN Evgrafova i A.A. Popovicha, SPb.: Izd-vo Politehn. un-ta, pp 439–449 (in Russian)
15. Lur'e AI (1961) Analytical mechanics. Nauka, Moscow, Gl. red. fiz._mat. lit, p 824 (in Russian)
16. Zhoga VV, Dyashkın-Titov VV, Nesmiyanov IA, Vorob'eva NS (2016) Manipulator of parallel-serial structure with a controlled gripper positioning task (Zadacha pozicionirovanija manipuljatora parallel'no-posledovatel'noj struktury s upravljaemym zahvatnym ustrojstvom). Mehatronika, avtomatizacija, upravlenie. No. 8. vol 17. pp 525–530 (in Russian)
17. Kolovskii MZ, Sloushch AV (1998) Foundations of industrial robot dynamics. Nauka, Moscow, Gl. red. fiz._mat. lit, p 240 (in Russian)
18. Korendesev AI, Salamandra BL, Tyves LI (2006) Theoretical basis of robotics. In 2 books. Institute of Machines Science named after A.A. Blagonravov of the Russian Academy of Sciences. Moscow: Nauka. Book 1, p 383 (in Russian)
19. Dyashkin-Titov VV, Vorob'eva NS, Terehov SE (2016) The algorithm of positioning the gripper of a manipulator-tripod (Algoritm pozicionirovanija zahvata manipuljatora-tripoda). Sovremennoe mashinostroenie: Nauka i obrazovanie: Materialy 5-j Mezhdunar. nauch.-prakt. konferencii/ Pod red. A.N. Evgrafova i A.A. Popovicha, SPb.: Izd-vo Politehn. un-ta, pp 634–644 (in Russian)

Simulation of the Dynamics of a Rotor on Foil Bearings

Vladimir V. Eliseev and Ekaterina A. Andriushchenko

Abstract The rotor rotation on foil bearings is considered with the tensor notation of the balance equations of momentum and angular momentum. Linearization of equations close to nominal mode follows to the linear eighth-order system with periodical coefficients. This system can be solved through computer mathematics technique (*Mathcad*).

Keywords Dynamics of a rigid rotor · Viscoelastic bearings · Tensor equations
Computer mathematics · Forced and parametric oscillations

Introduction

Consider the dynamics of a rigid rotor on foil bearings (see Fig. 1). In the limiting case of rigid bearings, the rotor would rotate about the fixed axis z of the Cartesian coordinate system x, y, z with the given angular velocity Ω.

A lot of books and articles [1–9] have been devoted to rotor dynamics. Authors have considered both rigid and elastically deformed rotors there, and for the anisotropy of bearings, dual bending stiffness, gravity force, temperature deformations, and so on. However, there is no widely accepted rotor theory.

The objective of this research is to present the theoretical foundations of the rigid rotors on foil bearings combined with computer simulation. The novelty is in the tensor calculus and computer simulation [10].

V. V. Eliseev · E. A. Andriushchenko (✉)
Peter the Great St. Petersburg Polytechnic University, Saint Petersburg, Russia
e-mail: katarina.and@yandex.ru

V. V. Eliseev
e-mail: yeliseyev@inbox.ru

© Springer International Publishing AG 2018
A. N. Evgrafov (ed.), *Advances in Mechanical Engineering*, Lecture Notes
in Mechanical Engineering, https://doi.org/10.1007/978-3-319-72929-9_6

Fig. 1 Scheme of rotor

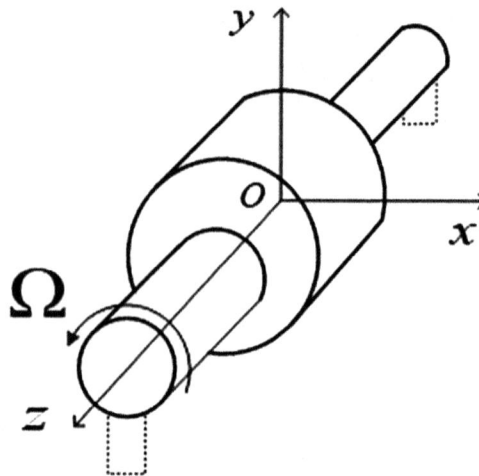

Derivation of a System of Ordinary Differential Equations

We describe the dynamics of a rigid rotor according to two vector ordinary differential equations (ODE). The first one expresses the balance of momentum or the principle of motion of the center of mass [11–13]:

$$m(\mathbf{r} + \ddot{\boldsymbol{\varepsilon}}) = \sum_j \mathbf{F}_j^e, \ \mathbf{r} = z_c \mathbf{k} + \mathbf{u}. \tag{1}$$

Here, m is the mass of the rotor, point means the time derivative, $\mathbf{r}$ is the position vector of the pole, $\mathbf{k}$ is the unit vector of axis z, $\mathbf{u}$ is the vector of displacement of the pole, and $\boldsymbol{\varepsilon}$ is the eccentricity vector. There is a sum of external forces (bearing reactions) on the right-hand side of the equation.

The second equation is the principle of balance of angular momentum:

$$(\mathbf{I} \cdot \dot{\boldsymbol{\omega}}) + m\boldsymbol{\varepsilon} \times \ddot{\mathbf{r}} = \sum_j \mathbf{M}_j^e. \tag{2}$$

Here, $\mathbf{I}$ is the inertia tensor of the rotor as a rigid body and $\boldsymbol{\omega}$ is the angular velocity vector. There is a sum of external moments on the right-hand side of the equation. In addition, there is the cross-link between Eqs. (1) and (2). It will disappear if the pole is the mass center ($\boldsymbol{\varepsilon} = 0$). The tensor calculus, which is used in this article, is outlined in the book [11].

The vectors $\mathbf{u}(t), \boldsymbol{\omega}(t)$ are the unknown functions. We may describe the angular orientation of the rotor by the rotation tensor $\mathbf{P}(t)$, if we integrate the equation $\dot{\mathbf{P}} = \boldsymbol{\omega} \times \mathbf{P}$ [11]. All expressions above are nonlinear. This is related, in particular, to the following:

$$\dot{\varepsilon} = \omega \times \varepsilon, \ \dot{\mathbf{I}} = \omega \times \mathbf{I} - \mathbf{I} \times \omega. \tag{3}$$

In accordance with theoretical mechanics [14, 15], there are no bearing reactions when the rotor is rotating about a fixed axis if two balancing conditions are fulfilled. Firstly, the center of mass has to be on the axis of rotation, meaning $\varepsilon = 0$ (static condition). Secondly, the axis of rotation has to be the principal axis of the tensor of inertia, meaning $\mathbf{I} \cdot \mathbf{k} = I_z \mathbf{k}$ (dynamic condition with the moment of inertia about axis z). Therefore, for the dynamics of the rotor on foil bearings, we take $\mathbf{I} = \mathbf{I}^0 + \mathbf{I}^1$, where the first summand satisfies the dynamic condition of balancing, and the second one is as small as the eccentricity ε.

Note that $\omega \neq \Omega \mathbf{k}$. The movement of the rotor is considered compound: the frame motion with velocity $\Omega \mathbf{k}$ and the relative motion with the vector of small rotation θ. In the case of relative motion, the unit coordinate vectors, eccentricity, inertia tensor, angular velocity and angular momentum get the small increments:

$$\tilde{\mathbf{e}}_j = \theta \times \mathbf{e}_j, \ \tilde{\varepsilon} = \theta \times \varepsilon, \ \tilde{\omega} = \dot{\theta} + \Omega \theta \times \mathbf{k},$$
$$\tilde{\mathbf{I}} = \theta \times \mathbf{I} - \mathbf{I} \times \theta, \ \widetilde{(\mathbf{I} \cdot \omega)} = \mathbf{I} \cdot \dot{\theta} + \theta \times \mathbf{I} \cdot \omega. \tag{4}$$

The derivation of this formula is in the book [11]. Note that the principal part of the tensor of inertia is

$$\mathbf{I}^0 = I_1 \mathbf{e}_1 \mathbf{e}_1 + I_2 \mathbf{e}_2 \mathbf{e}_2 + I_z \mathbf{k}\mathbf{k}. \tag{5}$$

This tensor rotates about the axis z with the velocity Ω (together with the unit vectors $\mathbf{e}_1, \mathbf{e}_2$). In the case of equal transverse moments of inertia $(I_1 = I_2)$, the tensor is constant, because $\mathbf{e}_1 \mathbf{e}_1 + \mathbf{e}_2 \mathbf{e}_2 = \mathbf{E}_\perp = \mathbf{E} - \mathbf{k}\mathbf{k}$—with the identity tensors in plane x, y and in space.

We take into account that ε, $\mathbf{I}^1$ are small and write the equations in variations in the vicinity of the nominal mode:

$$m(\ddot{\mathbf{u}} - \Omega^2 \varepsilon) = \sum \tilde{\mathbf{F}}^e_k$$
$$\mathbf{I}^0 \cdot \ddot{\theta} + \Omega \mathbf{I}^* \cdot \dot{\theta} = \sum \tilde{\mathbf{M}}^e_k, \tag{6}$$
$$\mathbf{I}^* \equiv \mathbf{k} \times \mathbf{I}^0 - \mathbf{I}^0 \times \mathbf{k} - I_z \mathbf{k} \times \mathbf{E}.$$

Note that the perturbation of $\mathbf{I}^1$ appears in the small quantities of second-order and is not present in the equations.

We determine forces and moments from the bearing reactions. The bearings are in the cross-sections $z = 0$, $z = l$ and have different elastic and dissipative characteristics along the horizontal and vertical:

$$z = 0: \mathbf{F}^0 = -\mathbf{c}^0 \cdot \mathbf{U}^0 - \mathbf{b}^0 \cdot \dot{\mathbf{U}}^0,$$
$$z = l: \mathbf{F}^l = -\mathbf{c}^l \cdot \mathbf{U}^l - \mathbf{b}^l \cdot \dot{\mathbf{U}}^l. \tag{7}$$

Here, we have four tensors describing the stiffness and viscosity of bearings; $\mathbf{U}, \dot{\mathbf{U}}$ are the displacements and velocities of the ends of the rotor:

$$\begin{aligned} \mathbf{U}^0 &= \mathbf{u} - z_c \boldsymbol{\theta} \times \mathbf{k}, \\ \mathbf{U}^1 &= \mathbf{u} + (l - z_c)\boldsymbol{\theta} \times \mathbf{k}. \end{aligned} \tag{8}$$

The relations for the velocities are similar. On the right-hand sides of Eq. (6), we obtain

$$\begin{aligned} \sum \mathbf{F}_i^e &= \mathbf{F}^0 + \mathbf{F}^l, \\ \sum \mathbf{M}_i^e &= -z_c \mathbf{k} \times \mathbf{F}^0 + (l - z_c)\mathbf{k} \times \mathbf{F}^l. \end{aligned} \tag{9}$$

Then, there is the second-order vector ODE system for functions $\mathbf{u}, \boldsymbol{\theta}$. We solve it for $\ddot{\boldsymbol{\theta}}$:

$$\begin{aligned} \ddot{\boldsymbol{\theta}} + \mathbf{R} \cdot \dot{\boldsymbol{\theta}} &= \mathbf{Q} \cdot \sum \mathbf{M}_i^e, \\ \mathbf{R} &\equiv \Omega \mathbf{Q} \cdot \mathbf{I}^*, \quad \mathbf{Q} \equiv (\mathbf{I}^0)^{-1}. \end{aligned} \tag{10}$$

The inverse tensor $\mathbf{Q} = (\mathbf{I}^0)^{-1}$ is the tensor $\mathbf{I}^0$ with the exchange $I_{1,2}$ to $I_{1,2}^{-1}$. We may keep only the perpendicular component in tensors (in plane x, y).

Computer Simulation

The ODE system in components has the eighth-order and is linear with periodical coefficients. It is possible to find the Cauchy problem solution with the help of computer mathematics. For this purpose, there are functions *rkfixed*, *Rkadapt* in *Mathcad* [10].

We transform the system of equations. Firstly, we rewrite the equations of displacement of the ends of the rotor:

$$\begin{aligned} \mathbf{U}^0 &= \mathbf{u} - z_c \Im \cdot \boldsymbol{\theta}, \\ \mathbf{U}^l &= \mathbf{u} + (l - z_c)\Im \cdot \boldsymbol{\theta}, \\ \Im &= -\mathbf{k} \times \mathbf{E} \end{aligned} \tag{11}$$

Secondly, we follow from tensors to matrices of components in basis x, y:

$$\begin{aligned} \sum F_i^e &= -c^+ u - b^+ \dot{u} - C\Im\theta - B\Im\dot{\theta} \\ \sum M_i^e &= \Im(Cu + B\dot{u} + D\Im\theta + G\Im\dot{\theta}) \end{aligned} \tag{12}$$

Here, we introduce the matrix notation:

$$\Im \equiv \begin{pmatrix} 0 & 1 \\ -1 & 0 \end{pmatrix}, c^+ \equiv c^0 + c^l, b^+ \equiv b^0 + b^l,$$

$$C \equiv -z_c c^0 + (l - z_c)c^l, B \equiv -z_c b^0 + (l - z_c)b^l,$$ (13)

$$D \equiv z_c^2 c^0 + (l - z_c)^2 c^l, G \equiv z_c^2 b^0 + (l - z_c)^2 b^l$$

The Levi-Chivita matrix $\Im$ allows for describing the cross-product.
We consider the tensors in (10) with the transition to the basis x, y:

$$\mathbf{I}^* = -\Im \cdot \mathbf{I}^0 + \mathbf{I}^0 \cdot \Im + I_z \Im \Rightarrow I^* = -\Im I^0 + I^0 \Im + I_z \Im, I^0 \equiv \begin{pmatrix} I_1 & 0 \\ 0 & I_2 \end{pmatrix},$$

$$Q = \alpha Q^0 \alpha^T, Q^0 \equiv \begin{pmatrix} I_1^{-1} & 0 \\ 0 & I_2^{-1} \end{pmatrix}, \alpha \equiv \begin{pmatrix} \cos \Omega t & -\sin \Omega t \\ \sin \Omega t & \cos \Omega t \end{pmatrix}, R = \Omega \alpha Q^0 I^* \alpha^T$$ (14)

Null denotes the components in the basis $\mathbf{e}_1, \mathbf{e}_2$. The time dependence is in the rotation matrix α only.

Now we represent the ODE system in the matrix form:

$$\dot{Y} = H(t)Y + h(t); Y \equiv \begin{pmatrix} u & \dot{u} & \theta & \dot{\theta} \end{pmatrix}^T;$$

$$H(t) \equiv \begin{pmatrix} O & E & O & O \\ -m^{-1}c^+ & -m^{-1}b^+ & -m^{-1}C\Im & -m^{-1}B\Im \\ O & O & O & E \\ Q(t)\Im C & Q(t)\Im B & Q(t)\Im D\Im & -R(t) + Q(t)\Im G\Im \end{pmatrix},$$ (15)

$$h(t) = \begin{pmatrix} o & \Omega^2 \varepsilon & o & o \end{pmatrix}^T$$

The coefficients include the sinusoid of frequency 2Ω; this is the parametric excitation. The presence of eccentricity in $h(t)$ gives us the harmonic excitation of frequency Ω that can cause forced oscillations.

In *Mathcad*, the calculation of the steel rotor was conducted with the parameters: the length is $l = 10$ m, the coordinates of the ends of sections (Fig. 1) are $z_1 = 2$, $z_2 = 6$ (m), and the radii are $r_1 = 0.3$, $r_2 = 0.7$, $r_3 = 0.4$ (m). Thus, the mass is $m = 6.81 \times 10^4$ kg, the coordinate of the mass center is $z_c = 4.73$ m, and the moments of inertia are $I_z = 1.32 \times 10^4$, $I_1 = 3.41 \times 10^5$, $I_2 = 3.07 \times 10^5$ (kg m²) (we reduced I_1 by 10% to study parametric resonance). The eccentricity is equal to $\varepsilon = 10^{-3}$ m. The properties of the supports are equal and isotropic, with the stiffness $c = 10^8$ N/m and viscous resistance coefficient $b = 10^6$ N s/m. The frequency is $\Omega = 55$ s^{-1}. This is close to the natural frequency $\sqrt{2c/m} = 54.2$ s^{-1}, but there is no resonance increase of amplitude.

Fig. 2 Displacements
$(m, \Omega = 55 \ s^{-1})$

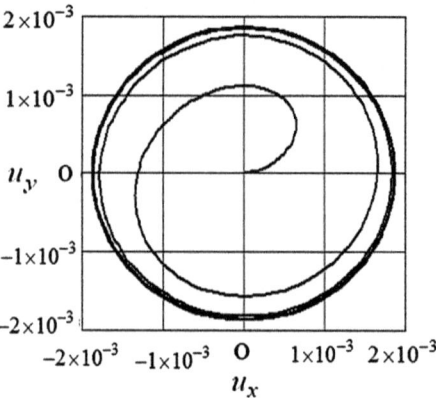

Fig. 3 Hodograph of angular velocity

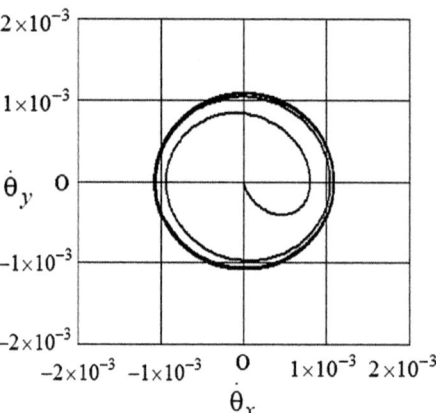

In Fig. 2, we show the hodograph of the displacement **u** with the initial conditions equal to zero. We can see that the regime with constant amplitude is set up. The hodograph of the angular velocity $\dot{\boldsymbol{\theta}}$ is presented in Fig. 3.

The calculation with the various angular velocities reveals the dependence of the average amplitude on the angular velocity $a(\Omega)$, which is shown in Fig. 4. We mediate by time T, using the following formula:

$$a(\Omega) = \frac{1}{(1 - \lambda)T} \int_{\lambda T}^{T} |\mathbf{u}(t, \Omega)| dt \tag{16}$$

Here, we introduce the small parameter λ to exclude the initial transient process and take $\lambda = 10^{-3}, T = 10\,\text{s}$.

Fig. 4 Dependence of average amplitude (mm) on angular velocity (s^{-1})

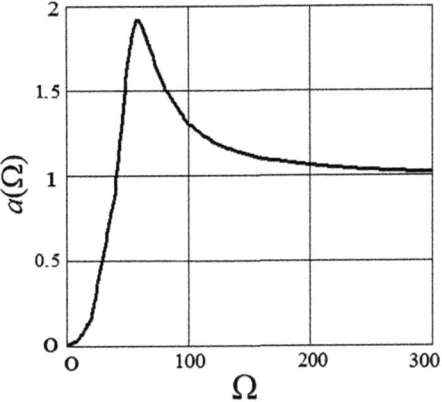

Fig. 5 Displacement when frequency is high

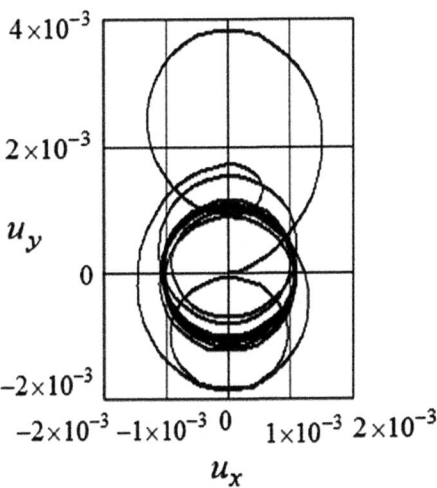

Such dependence is typical for forced oscillation with a load that is proportional to Ω^2. The more complicated effects of parametric excitation are not found.

However, it is worth noting that oscillations become more chaotic with the increase in frequency. In Fig. 5, we draw the hodograph of displacement when $\Omega = 200\ \text{s}^{-1}$.

Conclusion

In this research, we presented the novel equations of the dynamic theory of rigid rotors on foil bearings. We developed the technique of computer simulating in *Mathcad* based on this theory. The multivariate calculations carried out with this technique allow for choosing characteristics of bearings at the project stage and avoiding dangerous oscillations.

References

1. Gusarov AA (2004) Machine rotor balancing (Balansirovka rotorov mashin). Nauka, Moscow, p 267 (in Russian)
2. Genta G (2005) Dynamics of rotating systems. Springer, New York, p 658
3. Yamamoto T, Ishida Y (2001) Linear and nonlinear rotordynamics: a modern treatment with applications. Wiley, New York, p 348
4. Sperling L, Ryzhik B, Linz C, Duckstein H (2002) Simulation of two-plane automatic balancing of a rigid rotor. Math Comput Simul 58:351–365
5. Vance JM, Zeidan FY, Murphy B (2010) Machinery vibration and rotordynamics. Wiley, New York, p 412
6. Darlow MS (1989) Balancing of high-speed machinery. Springer, New York, p 184
7. Adams ML (2009) Rotating machinery vibration: from analysis to troubleshooting. CRC Press, Boca Raton, p 449
8. Lee CW (1993) Vibration analysis of rotors. Springer Science & Business Media, Dordrecht, p 314
9. Findeisen D (2013) System dynamics and mechanical vibrations: an introduction. Springer Science & Business Media, Dordrecht, p 383
10. Kiryanov D, Kiryanova E (2007) Computational science. Infinity Science Press, Hingham
11. Eliseev VV (2003) Mechanics of elastic bodies. St. Petersburg State Polytechnic University Publishing House, St. Petersburg, p 336 (in Russian)
12. Yeliseyev VV (2015) Dynamics, critical speeds and balancing of thermoelastic rotors In: Evgrafov A (ed) Advances in mechanical engineering. Lecture notes in mechanical engineering. Published by Springer International Publishing, Switzerland, pp 129–136. https://doi.org/10.1007/978-3-319-15684-2
13. Cheli F, Diana G (2015) Advanced dynamics of mechanical systems. Springer, Berlin, p 836
14. Menard KP (2008) Dynamic mechanical analysis: a practical introduction. CRC Press, Boca Raton, p 218
15. Pfeiffer F (2008) Mechanical system dynamics. Springer Science & Business Media, Dordrecht, p 575

Vibrations of Turbine Blades as Elastic Shells

Vladimir V. Eliseev and Artem A. Moskalets

Abstract We consider vibrations of short turbine blades which have the length and the width of the same order of magnitude. We use the modern variant of the classical Kirchhoff shells based on Lagrange mechanics. The equations in components are derived from the tensor equations with account for the natural twist of the blade. We take displacement approximation with coefficients that are generalized coordinates of the Lagrange equations. The algorithm of calculations with computer mathematics is proposed; it includes evaluation of the generalized masses and stiffnesses. We present a benchmark example with the shell modal analysis.

Keywords Turbine blades · Theory of shells · Direct tensor calculus
Equations in components · Displacement approximation · Computer mathematics
Lagrange equations · Modal analysis

Introduction

Turbine blades work in a heavy mode of forced oscillation under the action of steam or fluid jet pressure. Because dangerous oscillations with high amplitudes and unavoidable fatigue failure are possible, it is worth calculating both free and forced oscillations. Modal analysis with determination of the natural frequencies and natural modes at the design stage is of great importance for avoiding resonances.

A lot of books and articles [1–7] have been devoted to the vibrations of turbine blades. In the beginning, one-dimensional shaft models were considered [1, 2]. However, when the computing technologies of finite element analysis appeared, they started to dominate [8]. Nevertheless, it is not reasonable to use only

V. V. Eliseev · A. A. Moskalets (✉)
Peter the Great St. Petersburg Polytechnic University, Saint Petersburg, Russia
e-mail: artem.moskalec@gmail.com

V. V. Eliseev
e-mail: yeliseyev@inbox.ru

© Springer International Publishing AG 2018
A. N. Evgrafov (ed.), *Advances in Mechanical Engineering*, Lecture Notes
in Mechanical Engineering, https://doi.org/10.1007/978-3-319-72929-9_7

three-dimensional models with the finite element method. The modern one-dimensional models may be very efficient for the long blades [9–13].

In practice, the short blades occur; they have lengths and widths of the same magnitude. For this type of blade, it is natural to use the theory of shells. Great progress has been achieved in the theory of shells nowadays [14–18]. However, the complexity of the geometry of naturally twisted blades [19] causes problems in traditional calculation. The appearance of computer mathematics (Mathcad [20] and others) changed everything. It gives us the opportunity to solve complicated problems.

This paper is based on the modern theory of shells with direct tensor calculus [16], Lagrange's analytical mechanics [11, 16] and computer mathematics [20]. We present a benchmark example of calculation; more importantly, the algorithm that is proposed in this paper allows us to consider more complicated cases of shell geometry and forces (forced oscillations).

Information from the Theory of Shells

We show the blade schematically in Fig. 1. The axis z passes through the cross-sectional centers of mass. The shapes of cross-sections for all values of z are the same, but they are rotated by an angle αz, where α is the parameter of twist. The axes x, y (with the unit vectors $\mathbf{i}$, $\mathbf{j}$) are fixed, the axes ξ, η (with the unit vectors $\mathbf{e}_\xi, \mathbf{e}_\eta$) are rotated together with the cross-section. The cross-section of the blade as a shell is the curve specified by the relation $\eta(\xi)$, $\xi_0 \leq \xi \leq \xi_1$.

In the theory of shells, we first specify the geometry of the shell as a material surface. The position vector is a function of two curvilinear coordinates $\mathbf{r}(\gamma^1, \gamma^2)$. Each point lies on the concurrence of two coordinate lines. Vectors of derivatives

Fig. 1 Blade geometry

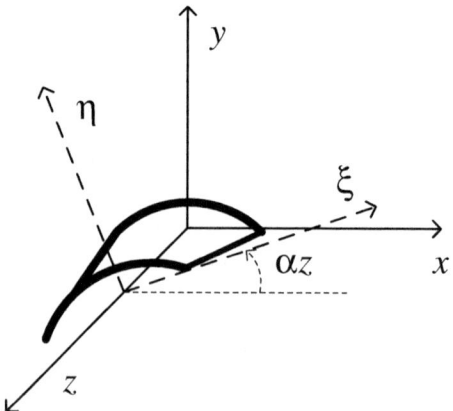

$\mathbf{r}_\beta = \partial \mathbf{r}/\partial \gamma^\beta \equiv \partial_\beta \mathbf{r}$ are tangential to these two coordinate lines. The unit normal vector reads as

$$\mathbf{n} = m^{-1}\mathbf{r}_1 \times \mathbf{r}_2, \quad m \equiv |\mathbf{r}_1 \times \mathbf{r}_2|. \tag{1}$$

The vectors $\mathbf{r}_\beta$ ($\beta = 1, 2$) form the basis in the tangent plane. It is possible to decompose an arbitrary vector in space in the following way: $\mathbf{u} = u^\beta \mathbf{r}_\beta + u_n \mathbf{n}$. From now on, we will use the summation rule over the indices of different levels. The coefficients u^β are called "contravariant components". Also, we introduce the reciprocal basis so that the equality $\mathbf{r}_\alpha \cdot \mathbf{r}^\beta = \delta_\alpha^\beta$ is satisfied (δ_α^β is the Kronecker delta, which equals one for $\alpha = \beta$ and zero for $\alpha \neq \beta$). Hence, $u^\beta = \mathbf{u} \cdot \mathbf{r}^\beta$. A vector may also be expanded with respect to the co-basis with the contravariant components: $u^\beta \mathbf{r}_\beta = u_\beta \mathbf{r}^\beta$. Now we can introduce the Hamilton operator on the surface:

$$\nabla = \mathbf{r}^\beta \partial_\beta; \quad d\phi(\gamma^\beta) = d\mathbf{r} \cdot \nabla \phi. \tag{2}$$

The expression for ∇ is constructed such that the equality for $d\phi$ is fulfilled. The first and second metric tensors on the surface are commonly introduced:

$$\mathbf{a} = \nabla \mathbf{r} = \mathbf{r}^\beta \mathbf{r}_\beta = a_{\alpha\beta}\mathbf{r}^\beta \mathbf{r}^\beta = a^{\alpha\beta}\mathbf{r}_\alpha \mathbf{r}_\beta, \quad \mathbf{b} = -\nabla \mathbf{n}. \tag{3}$$

They are symmetric. $\mathbf{a} = \mathbf{E} - \mathbf{nn}$ is the identity tensor in the tangent plane. Its covariant components $a_{\alpha\beta}$ determine lengths and angles on the surface. The tensor $\mathbf{b}$ prescribes the surface curvature. A lot of books have been devoted to the surface geometry and tensor calculus application; however, to present the basics of the shell theory in this paper, we only need just some of their essentials.

We may describe the classic Kirchhoff shells consisting of "material normal", which are the objects with five degrees of freedom. For each particle, we assign the deflection $\mathbf{u}$, as well as the vector rotation characteristic $\tilde{\mathbf{n}} = \boldsymbol{\varphi}$; the symbol " $\sim$ " means small variation during deformation. Now we can derive a formula:

$$\boldsymbol{\varphi} = -\nabla \mathbf{u} \cdot \mathbf{n}. \tag{4}$$

We took into account that $\mathbf{n}$ remains to be normal. Formula (4) corresponds to the model without shear. Relation (4) is not valid in the case of the more complicated Timoshenko model, with account for shear.

The two following strain tensors play the most important role:

$$\boldsymbol{\varepsilon} = (\nabla \mathbf{u})_\perp^S, \quad \boldsymbol{\kappa} = -(\nabla \boldsymbol{\varphi})_\perp + \mathbf{b} \cdot \nabla \mathbf{u}^T. \tag{5}$$

They are symmetric and in the tangent plane.

The denotation $\perp$ means the components in the tangent plane, $(\ldots)^S, (\ldots)^T$, denote symmetrization and transposition, respectively. The tensor $\boldsymbol{\varepsilon}$ determines the stretch and shear in the tangent plane. $\varepsilon_{\alpha\beta} = \widetilde{a}_{\alpha\beta}/2$ are the components of $\boldsymbol{\varepsilon}$. The second tensor $\boldsymbol{\kappa}$ defines the bending and torsional strains; $\kappa_{\alpha\beta} = \widetilde{b}_{\alpha\beta}$.

The tensors $\boldsymbol{\varepsilon}$, $\boldsymbol{\kappa}$ are called "strain tensors", because six functions $a_{\alpha\beta}(\gamma)$, $b_{\alpha\beta}(\gamma)$ of coordinates γ^{α} determine the form of the surface up to rigid-body displacements, so the deformation is determined by the variation of these functions.

The potential energy of deformation distributed per unit area reads as

$$\widehat{\Pi}(\boldsymbol{\varepsilon}, \boldsymbol{\kappa}) = \frac{Eh}{2(1 - \nu^2)} \left[\nu\varepsilon^2 + (1 - \nu)\boldsymbol{\varepsilon} \cdot \cdot \boldsymbol{\varepsilon} + \frac{h^2}{12} \left(\nu\kappa^2 + (1 - \nu)\boldsymbol{\kappa} \cdot \cdot \boldsymbol{\kappa} \right) \right]. \quad (6)$$

Here, E is the Young modulus, ν is the Poisson ratio, h is the thickness, and ε, κ are the first invariants (traces) of the corresponding tensors. The expression for $\widehat{\Pi}(\boldsymbol{\varepsilon}, \boldsymbol{\kappa})$ is the same as in the theory of plates [16].

Differentiating (6), we may derive the expressions for the force and moment tensors. Then, we evaluate stresses using the formulae of the theory of plates for the plane stress state and bending [16].

However, it is not necessary to use the balance equations, because we will employ the Lagrange equations:

$$\left(\frac{\partial K}{\partial \dot{q}_i} \right)^{\bullet} - \frac{\partial K}{\partial q_i} = -\frac{\partial \widehat{\Pi}}{\partial q_i} + Q_i, \ \Pi = \int \widehat{\Pi} do, \ K = \int \rho h |\dot{\mathbf{u}}|^2 do \quad (7)$$

(here $\Pi = \int \widehat{\Pi} do$ is the potential energy, $K = \int \rho h |\dot{\mathbf{u}}|^2 do$ is the kinetic energy, and ρ is the density). $do = m \, d\gamma^1 d\gamma^2$ is the unit area on the surface. Before using Eqs. (7), we should introduce the generalized coordinates q_i, as well as construct the expressions for $\Pi(q_i)$, $K(q_i, \dot{q}_i)$ and for the generalized forces Q_i.

Shape of Shell and Computer Modeling

We refer to Fig. 1 and prescribe the surface by the equation

$$\mathbf{r}(z, \xi) = z\mathbf{k} + \mathbf{x}, \ \mathbf{x} \equiv \xi \mathbf{e}_{\xi}(z) + \eta(\xi)\mathbf{e}_{\eta}(z) \quad (8)$$

In order to construct the basis, we account for the derivation formulae $\mathbf{e}'_{\xi}(z) = \alpha\mathbf{e}_{\eta}$, $\mathbf{e}'_{\eta}(z) = -\alpha\mathbf{e}_{\xi}$:

$$\mathbf{r}_1 = \partial_z \mathbf{r} = \mathbf{k} + \alpha(\xi \mathbf{e}_{\eta} - \eta \mathbf{e}_{\xi}), \ \mathbf{r}_2 = \partial_{\xi} \mathbf{r} = \mathbf{e}_{\xi} + \eta'(\xi)\mathbf{e}_{\eta} \quad (9)$$

Then, we find

$$\mathbf{n} = m^{-1}(\mathbf{e}_\eta - \eta'\mathbf{e}_\xi - \alpha(\xi + \eta\eta')\mathbf{k}), \; m = \sqrt{1 + \eta'^2 + \alpha^2(\xi + \eta\eta')^2},$$
$$\nabla = m^{-1}\mathbf{n} \times (\mathbf{r}_1\partial_\xi - \mathbf{r}_2\partial_z). \tag{10}$$

We approximate the displacements of the shell as follows:

$$\mathbf{u} = \mathbf{U}(z,t) - \mathbf{U}' \cdot \mathbf{x} + s(z,t)(\xi - \xi_0)(\xi_1 - \xi)\mathbf{e}_\eta \tag{11}$$

The first two terms correspond to the elementary beam theory with the deflection vector $\mathbf{U}$. The third term defines the deformation of the cross-section with the generalized coordinate s in a simple manner (by square law). This approximation allows us to use equations with one coordinate (as in the rod theory) instead of two.

Let us approximate the introduced functions as well:

$$u_x(z,t) = \sum_{i=1}^{N} U_{xi}(t)\Phi_i(z) = U_x^T\Phi, u_y(z,t) = U_y^T\Phi,$$
$$s(z,t) = \sum_{i=1}^{N_s} S_i(t)\Psi_i(z) = S^T\Psi, \tag{12}$$

where the coordinate functions Φ_i, Ψ_i must satisfy the boundary conditions.

Equations (12) are the matrix equations with the columns U_x, U_y, S. $\Phi(0) = \Phi'(0) = \Psi(0) = 0$ because of the solid support at $z = 0$. We take the power dependencies:

$$\Phi_1 = z^2, \; \Phi_2 = z^3, \ldots, \; \Psi_1 = z, \; \Psi_2 = z^2, \ldots \tag{13}$$

For the blades as beams, such approximation was successful [9, 12].

With approximation, the energy of the shell is represented by the quadratic form of $\mathbf{U}$:

$$2\Pi = U_x^T C_{xx} U_x + U_y^T C_{yy} U_y + 2U_x^T C_{xy} U_y$$
$$+ S^T C_{ss} S + 2U_x^T C_{xs} S + 2U_y^T C_{ys} S,$$
$$2K = \dot{U}_x^T M_{xx} \dot{U}_x + \dot{U}_y^T M_{yy} \dot{U}_y + 2\dot{U}_x^T M_{xy} \dot{U}_y$$
$$+ \dot{S}^T M_{ss} \dot{S} + 2\dot{U}_x^T M_{xs} \dot{S} + 2\dot{U}_y^T M_{ys} \dot{S}. \tag{14}$$

The stiffness and inertia matrices are the corresponding integrals.

For the purpose of illustration, we take $N = N_s = 1$ so that all of the matrices and columns of (14) are numbers. By setting $U_x = 1$, $U_y = S = 0$, we can find C_{xx}, M_{xx}, etc. In the case of free vibrations, we have the following ODE system:

$$M_{xx}\ddot{U}_x + M_{xy}\ddot{U}_y + M_{xs}\ddot{S} + C_{xx}U_x + C_{xy}U_y + C_{xs}S = 0,$$
$$M_{xy}\ddot{U}_x + M_{yy}\ddot{U}_y + M_{ys}\ddot{S} + C_{xy}U_x + C_{yy}U_y + C_{ys}S = 0, \tag{15}$$
$$M_{xs}\ddot{U}_x + M_{ys}\ddot{U}_y + M_{ss}\ddot{S} + C_{xs}U_x + C_{sy}U_y + C_{ss}S = 0.$$

Now we can perform the modal analysis. By substituting $-\omega^2 U_x$ instead of $\ddot{U}_x$, we have the generalized eigenvalue problem (ω is the normal frequency):

$$(C - \omega^2 M)\Phi = 0, \tag{16}$$

which is solved by means of Mathcad [20].

In the case of forced vibrations, the Lagrange equations read as

$$M\ddot{U} + CU = Q(t). \tag{17}$$

The column of loads $Q(t)$ is calculated according to the virtual work $\delta A = Q^T \delta U$.

In the benchmark example, we considered the steel blade of the straight helicoid shape with the length $L = 20$ cm, width $2L = 20$ cm, thickness $h = 1$ cm and parameter of twist $\alpha = 1$ m^{-1}.

The stiffness and inertia matrices are calculated in Mathcad:

$$C = \begin{pmatrix} 3.44 \times 10^5 & 3.44 \times 10^4 & -713 \\ 3.44 \times 10^4 & 4.69 \times 10^3 & -72.2 \\ -713 & -72.2 & 26.3 \end{pmatrix},$$
$$M = \begin{pmatrix} 4.63 \times 10^{-4} & 2.46 \times 10^{-5} & -1.98 \times 10^{-6} \\ 2.46 \times 10^{-5} & 3.04 \times 10^{-4} & 1.23 \times 10^{-5} \\ -1.98 \times 10^{-6} & 1.23 \times 10^{-5} & -6.66 \times 10^{-7} \end{pmatrix}. \tag{18}$$

Solving the generalized eigenvalue problem, we get the eigenfrequencies and eigenmodes:

$$\omega_1 = 2.731 \times 10^4, \quad \omega_2 = 1.94 \times 10^3, \quad \omega_3 = 1.355 \times 10^4;$$
$$\Phi_1 = (-0.882 \quad -0.022 \quad -1),$$
$$\Phi_2 = (0.045 \quad -0.469 \quad -1), \tag{19}$$
$$\Phi_3 = (6.2 \times 10^{-3} \quad -0.042 \quad 1).$$

It can be seen that the second mode corresponds to the oscillations in the direction of the minimal stiffness. In the first and third modes, the deformation in the cross-sectional plane is more substantial.

Conclusion

In the present work, we considered the vibrations of short blades. The components of strain tensors were derived from the tensor equations of the general theory of shells. We calculated the strain energy using these components. We proposed the Lagrange equations, in which the coefficients of displacement approximation played the role of generalized coordinates. The numerical and symbolic algorithms of Mathcad were used. We presented a benchmark example of the modal analysis of a helicoid blade.

References

1. Birger IA, Shorr BF et al (1981) Dynamics of aviation gas turbine engines (Dinamika aviatsionnykh gazoturbinnykh dvigatelei). Mashinostroyenie, Moscow, p 232 (in Russian)
2. Levin AV, Borishanskiy KN, Konson ED (1981) Strength and vibration of blades and disks of steam turbines (Prochnost i vibratsiya lopatok i diskov parovyih turbin). Mashinostroyenie, Leningrad, p 710 (in Russian)
3. Bloch HP, Singh MP (2009) Steam turbines: design, applications and rerating. McGraw-Hill, New York, p 433
4. Leyzerovich AS (2005) Wet-steam turbines for nuclear power plants. PennWell, USA, p 481
5. Ilchenko BV, Gizzatullin RZ, Yarullin RR (2011) Stress-strain fields for blades of turbine K-210-130 under operation loading. Treatises Akademenergo (Trudy Akademenergo) 3: 74–81 (in Russian)
6. Magerramova LA (2013) Increasing resource of the aviation gas turbine blades by calculation methods. Herald Mosc Aviat Inst (Vestnik Moskovskogo Aviatsionnogo Instituta) 20(1): 58–70 (in Russian)
7. Melnikova GV, Shorr BF, Salnikov AV, Nigmatullin RZ (2014) Automated dynamic optimization of the blades of the gas turbine engines. Herald Mosc Aviat Inst (Vestnik Moskovskogo Aviatsionnogo Instituta) 21(1):76–85 (in Russian)
8. Leonov VP, Schastlivaya IA, Igolkina TN et al (2014) Application of finite element method for simulation of stress-strain state in manufacturing of long turbine blades made of high-strength titanium alloys. Inorg Mat Appl Res 5(6):578–586. https://doi.org/10.1134/S2075113314060069
9. Eliseev VV, Moskalets AA (2014) Vibrations of turbine blades as naturally twisted rod. In: Modern mechanical engineering. Science and education. Materials of the international scientific conference, Publishing house of the Polytechnic University, SPb., vol 4, pp 344–350
10. Eliseev VV, Moskalets AA, Oborin EA (2015) Deformation analysis of turbine blades based on complete one-dimensional model. Russ J Heavy Mach 5:35–38 (in Russian)
11. Eliseev VV, Moskalets AA, Oborin EA (2015) Applying of Lagrange equations to calculation of turbine blade vibration. Handbook. An Eng J 8:21–24. https://doi.org/10.14489/hb.2015.08 (in Russian)
12. Eliseev VV, Moskalets AA (2017) Computational technique of plotting campbell diagrams for turbine blades. Advances in mechanical engineering. Lecture Notes in Mechanical Engineering. Springer, pp 37–44. https://doi.org/10.1007/978-3-319-53363-6
13. Eliseev VV, Moskalets AA, Oborin EA (2016) One-dimensional models in turbine blades. Advances in mechanical engineering, Lecture notes in mechanical engineering, Springer, pp 93–104. https://doi.org/10.1007/978-3-319-29579-4

14. Pietraszkiewicz W (1988) Geometrically non-linear theories of thin elastic shells. Institut für Mechanik der Ruhr-Universität Bochum
15. Novozhilov VV, Chernykh KF, Mikhajlovskij EM (1991) Linear theory of thin shells (Linejnaya teoriya tonkih obolochek). Politekhnika, Leningrad, p 656 (in Russian)
16. Eliseev VV (2003) Mechanics of elastic bodies (Mekhanika uprugih tel). St. Petersburg State Polytechnic University Publishing House, St. Petersburg, p 336 (in Russian)
17. Eremeev VA, Zubov LM (2008) Mechanics of elastic shells (Mekhanika uprugih obolochek). Nauka, Moscow, p 280 (in Russian)
18. Eliseev VV, Vetyukov YM (2010) Finite deformation of thin shells in the context of analytical mechanics of material surfaces. Acta Mech 209:43–57. https://doi.org/10.1007/s00707-009-0154-7
19. Eliseev VV, Vetyukov YM, Zinov'eva TV (2011) Divergence of a helicoidal shell in a pipe with a flowing fluid. J Appl Mech Tech Phys 52:450–458. https://doi.org/10.1134/S0021894411030151
20. Kiryanov D, Kiryanova E (2007) Computational science. Infinity Science Press, Hingham

Contact Forces Between Wheels and Railway Determining in Dynamic Analysis. Numerical Simulation

Kirill V. Eliseev

Abstract Systems that include instrumented wheelsets and algorithms of measurement evaluation are used to determine wheel-rail contact forces while a train is moving. Here, some results of numerical testing of a developed algorithm in dynamic analysis are presented. All of the stages of data evaluation that were realized will be used for full-scale experiments.

Keywords Structural mechanics · Strain measurement · Simulation
Contact forces · Railway technology · Dynamics · Filtering

Introduction

To obtain parameters of contact interaction between wheel and rail, an algorithm has been developed that allows us to evaluate forces and contact coordinates based on measured strains. The measurement system consists of

- an instrumented wheelset with installed strain gauges;
- a system of data transfer and recording;
- an algorithm of data processing.

During development the task of alignment of gauges has been solved [1]. A scheme with two measurement circles (MC), an inner and an outer (MC1 and MC2), with 16 radial strains gauges on each, has been chosen (Fig. 1). Thus, we have measurement radii (MR) 1–16 and diameters (MD) 1–8. The radius of gauges 1 and 17 is IR1.

We establish the relation between forces R (3 components) and measured strains ε (32 components) for one wheel as

$$R = B\varepsilon, \tag{1}$$

K. V. Eliseev (✉)
Peter the Great St. Petersburg Polytechnic University, Saint Petersburg, Russia
e-mail: kir.eliseev@gmail.com

© Springer International Publishing AG 2018
A. N. Evgrafov (ed.), *Advances in Mechanical Engineering*, Lecture Notes
in Mechanical Engineering, https://doi.org/10.1007/978-3-319-72929-9_8

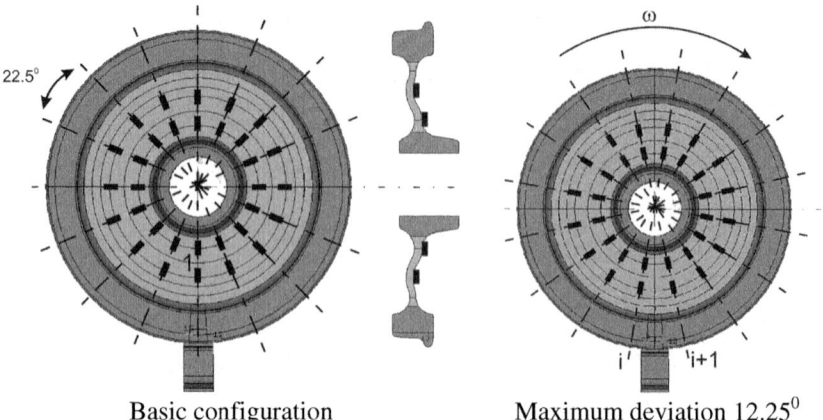

Basic configuration   Maximum deviation 12.25^0

Fig. 1 Gauges on wheel disk

where the B matrix was initially obtained in a numerical model and can be adjusted during calibration on the test-rig [2]. In general, this matrix depends on the contact position. The algorithm of contact coordinate evaluation is described in [3].

Accounting for the Wheelset Angle of Rotation

One fixed wheelset orientation was used while the initial investigation of the scheme of force evaluation was carried out. The basic configuration is the one in which the first measurement radius MR1 is vertical and these gauges are near contact (Fig. 1).

Relations between measured strains and contact forces were obtained for the basic configuration. It is obvious that all of the configurations that are rotated by 1/16 of the full revolution can be achieved by renumbering the virtual gauges. In a numerical model, relations do not change due to such rotation.

The wheelset rotation angle must be taken into consideration. There is no guarantee that in the time moment of strain measurement, there exists a measurement diameter that has a vertical position. (In patent [4] and some other articles measurement schemes for specific wheel orientation is suggested, synchronization equipment is used.)

The following algorithm is proposed. Cubic interpolation periodic splines [5] are built based on 16 strain values on each measurement circle. Analysis of the splines' local extreme values can suggest the angle of rotation. This procedure was tested for static analysis [1]. Then, these splines can be used to predict strains at locations that correspond to the basic configuration.

Wheelset Dynamics Numerical Modeling

Three numerical models of one wheel were used to check algorithm quality. All models use the finite-element method and were built in the ANSYS program [6]. Models were simplified in comparison with the static model—first-order elements with linear displacement interpolation are used in the measurement region instead of second-order, and there is no mesh refinement in the contact area. The purpose is analysis time and results file size reduction. Finite-element meshes consist of hexahedra and prism elements; there are 160 elements on any circle. A wheel contains approximately 100 thousand nodes and 120 thousand elements.

The Ls-Dyna program was used for dynamics analysis in [7, 8]. This program uses an explicit time integration scheme, while ANSYS uses an implicit one [6]. The explicit scheme needs a small time increment and is preferable for short period analyses like crash and explosion; a large amount of elements must be used.

Model 1 consists of one wheel on a rail (Fig. 2). Contact conditions are introduced. Axis movement with desired speed and time-dependent vertical force are defined. It is difficult to define the proper lateral and longitudinal forces, but corresponding reactions exist due to movement. Model 1 is the closest to a real wheel on a rail.

Models 2 and 3 consist of standalone wheels. Contact with the rail is modeled by definition of the point load (3 components) at some node of the model.

In Model 2, the wheel is rotating at a specified speed. The loading history consists of steps: every time step corresponds to a rotation of 1/160 of a full circle. Load continuity is obtained via linear interpolation of loads for neighbor time steps (Figs. 3 and 4).

Analysis of Model 3 is performed in rotation with a specified velocity frame. To model the loading history, point forces are applied onto the neighbor node at each time step. In [9], element pressure is applied in the same manner.

Fig. 2 Model 1 finite element mesh

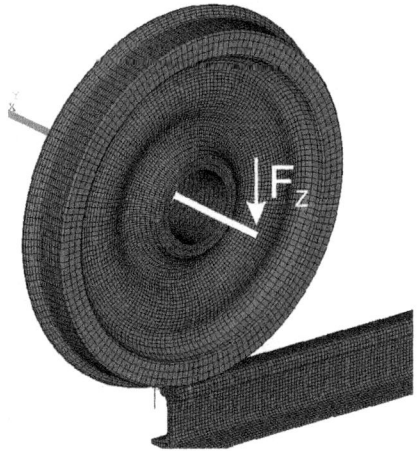

Fig. 3 Forces versus time
linear interpolation

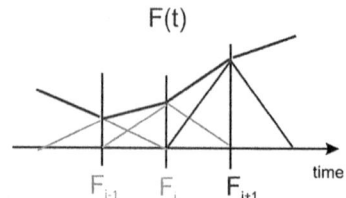

Fig. 4 Force application for
Models 2 and 3

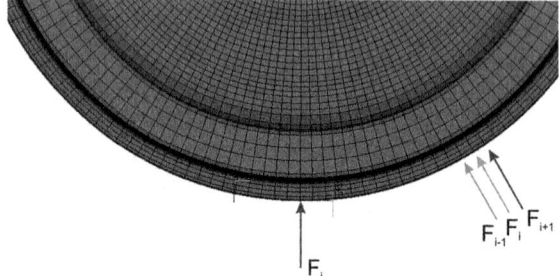

Initial Conditions and Damping

Initial conditions must be defined for Models 1 and 2 that correspond to the wheel movement with the desired speed.

Static analysis of a wheel under centrifugal loads can be performed, and then displacements in all nodes can be evaluated, thus we have initial displacements. Initial velocities in all nodes can be evaluated analytically. Initial velocities in Model 1 contain a translational velocity along the railway. Model 3 does not need initial conditions.

To reduce initial oscillations, forces were applied with linear growth at some initial steps from zero to prescribed values. Nevertheless, results for Models 1 and 2 had significant oscillations of stresses and strains at some of the first steps. The possible reason for this are features of the numerical time integration scheme.

To reduce the resulting oscillations, viscous material damping is usually introduced. In finite element analysis, a system of equations for a vector of model displacements U has the form

$$M\ddot{U} + C\dot{U} + KU = F(t), \quad C = \alpha M + \beta K. \tag{2}$$

Here, M and K are the mass and stiffness matrices and C is the damping matrix according to Rayleigh with coefficients α and β. These coefficients define damping relative to the oscillation frequency, the former being responsible for lower frequencies, the latter for higher. They depend on structure and can be obtained from experiments for a working range of frequencies.

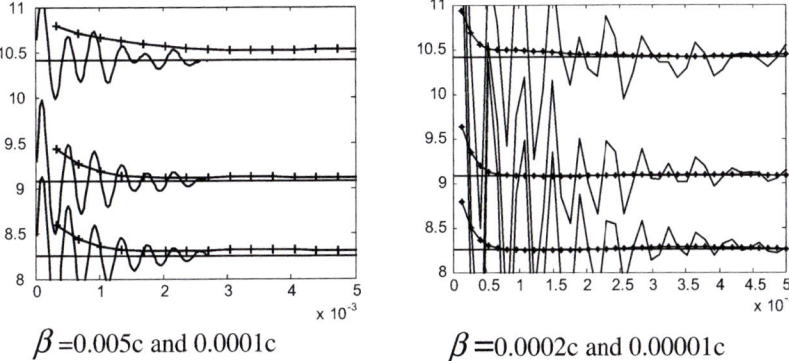

β =0.005c and 0.0001c β =0.0002c and 0.00001c

Fig. 5 Stresses versus time for different values of β relative to "static solution"

Here, coefficient β was selected as follows (compare with [10]). Analysis of the rotating wheel is solved as a dynamic analysis with initial conditions from the corresponding "static analysis" under centrifugal loads. Results are compared. It is expected that after a certain period of time, the dynamic solution must be close to the static one. Figure 5 shows a comparison of stresses at certain points of the wheel, the horizontal lines corresponding to the "static solution" that does not depend on time.

All variants, even those with inadequate damping, give us the right displacements and velocities vs. time dependencies, and the stress distribution has axial symmetry. However, value $\beta = 0.005c$ gives us fast decay but the wrong results, and $\beta = 0.00001c$ gives us insufficient damping. The next analysis variants use values 0.0001c and 0.0002c for β.

The value of 0.0001c was validated on the test-rig for the first natural frequency 140 Hz. The vibrations of the wheels were excited and the strain measurements were analyzed (Fig. 6).

Fig. 6 Typical strain decay on a test-rig

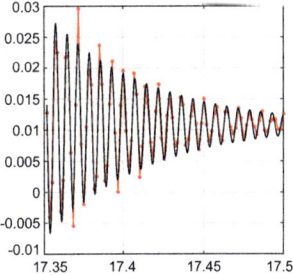

Results Post-processing

In the case of a motionless wheel in static analysis, we can use strains both in Cartesian and cylindrical coordinate systems. There are no difficulties working with strain tensor components. The same is true for Model 3.

In the case of large displacements, including large rotations, results on the screen and in the database are generated differently by the ANSYS program.

Some results for a rotating wheel disk under a vertical force at the lower wheel point are presented in Figs. 7 and 8, showing stress distribution in a cylindrical coordinate system. The wheel rim sector is shown for orientation purposes only. We expect symmetric stress distribution due to symmetric loading. But we can see the expected distribution only for ration angles that are multiples of 180°.

Visualization in a Cartesian coordinate system also cannot help in analyzing the results.

While analyzing strains evaluated using a spline for the base configuration, it was discovered that the results near the contact have a high degree of errors, especially on the outer circle IC2. This is due to the very variable values of strains

Fig. 7 Radial stress for
rotation angle 180° and 360°

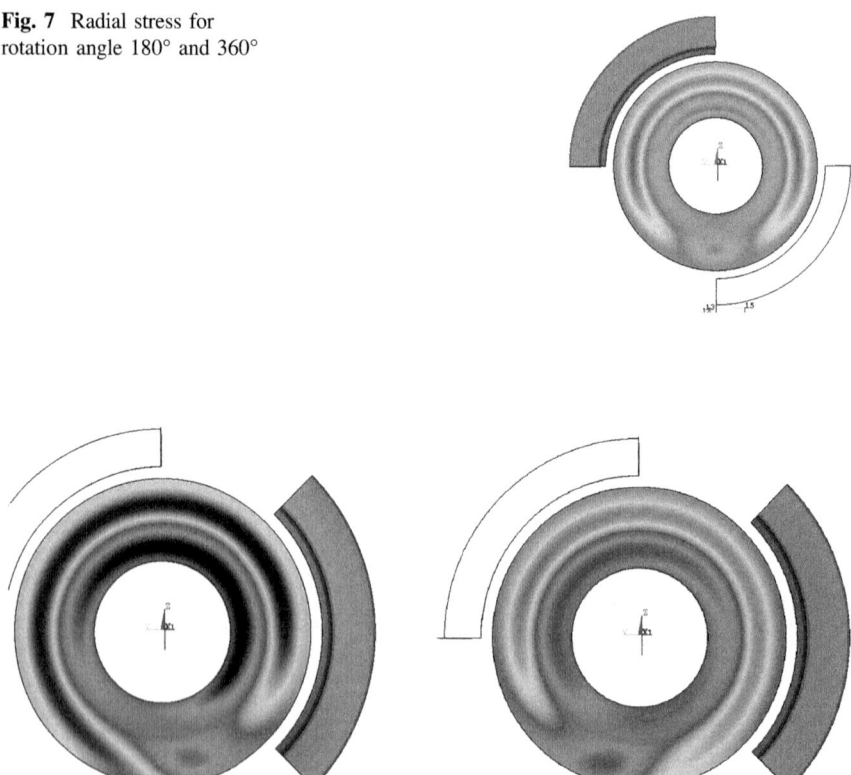

Fig. 8 Radial and circumferential stress for rotation angle 135°

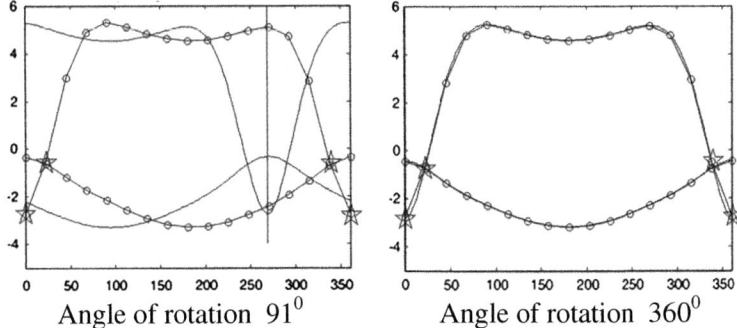

Fig. 9 Radial strain distribution on two circles: o—obtained via virtual gauges, others—evaluated for base configuration

near the contact; small errors in the spline approximation lead to big errors in the strains. One possible method to deal with these errors is not to use strains near the contact area (see the star points, Fig. 9).

Results for Model 2

Forces obtained with a simplified train model [11] are applied to the model for some intervals of time. The train velocity is 100 km/h. The solution does not have a high frequency oscillation of strains, and signal filtering is not used. Figures 10 and 11 present some results of the restored forces.

The results obtained using gauges near the contact are pulsating due to changing spline errors.

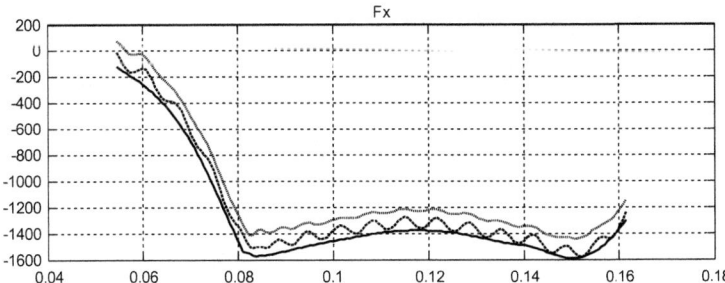

Fig. 10 Lateral force, N. Solid line—applied. Dashed and dotted—restored using different amounts of gauges

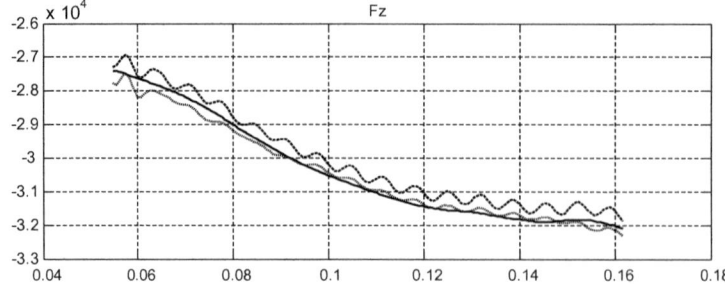

Fig. 11 Vertical force, N. Solid line—applied. Dashed and dotted—restored using different amounts of gauges

Results for Model 1

The next results are for a moving (100 km/h) wheel with a vertical load 50 kN applied at the bearing.

The virtual gauge strains have high frequency oscillations, and a low frequency filter is used (see Fig. 12).

The wheel's angle of rotation is usually restored with good quality (Fig. 13). There are some variations relative to the applied velocity due to some initial vibrations.

Figures 14 and 15 present gauge forces obtained from finite-element analysis and restored via strain. It can be stated that the number of errors is low in comparison to the applied vertical load level.

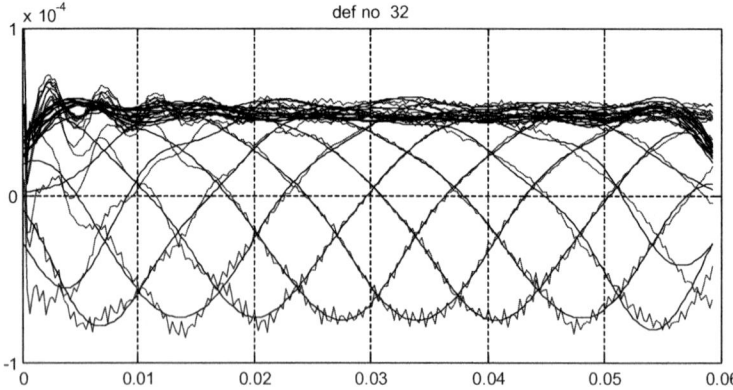

Fig. 12 Virtual gauges strains for MC2. Dotted line—model results, solid—strains after filtering

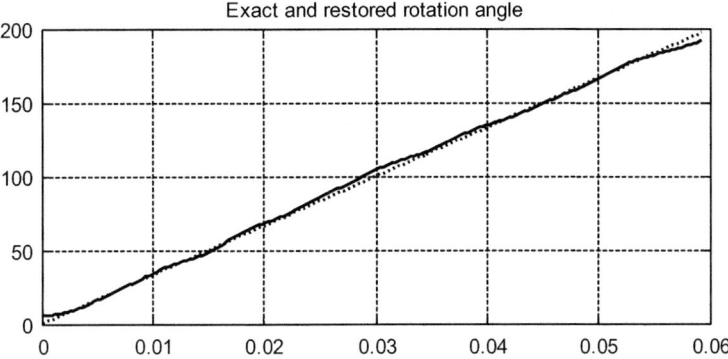

Fig. 13 Defined (dotted) and evaluated (solid) rotation angle (°) versus time, s

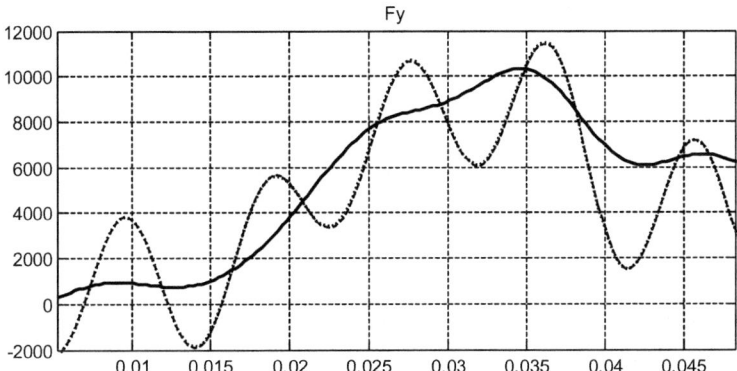

Fig. 14 Longitudinal force, N. Exact load after filtering (solid). Dashed—restored

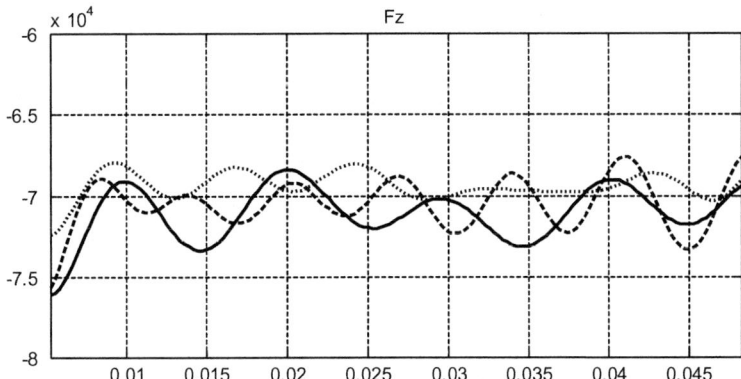

Fig. 15 Vertical force, N. Exact load after filtering (solid). Dashed and dotted—restored using different amounts of gauges

Conclusion

This work is dedicated to the practical realization of a contact force evaluation scheme during wagon movement by means of numerical models. The main stages are explained, and some problems are mentioned, along with their solutions. Acceptable results were obtained in numerical simulations.

References

1. Eliseev K, Ispolov I, Orlova A (2013) Contact forces between wheelset and rails determining. St. Petersburg State Polytechnical Univ J 4–1(183):262–270 (rus)
2. Eliseev K, Migrov A, Orlova A (2012) Design of test-rig for the calibration of instrumented wheelsets (rus). Transp Probl 2012, 474–479 (rus) Silesian University of Technology Faculty of Transport
3. Eliseev K (2017) Wheelsets and railways. Determining contact-points coordinates. In: Evgrafov A (ed) Advances in mechanical engineering. Lecture notes in mechanical engineering. Springer International Publishing, Switzerland, pp 177–187
4. Krasnov O et al (2010) Patent no 2441206 RU. 02.11.2010 Device for measuring of vertical and lateral forces between wheel and rail. https://patents.google.com/patent/RU2441206C1/en
5. Zavialov JS, Kvasov BI, Miroshnichenko VL (1980) Methods of spline – functions. Moscow, Nauka, 352 p. (rus)
6. ANSYS Release 15 user guide, 2014
7. Cui X et al (2015) Effect of the wheel/rail contact angle and the direction of the saturated creep force on rail corrugation. Wear 330–331:554–562
8. Vo K et al (2015) FE method to predict damage formation on curved track for various worn status of wheel/rail profiles. Wear 322–323:61–75
9. Ronasi H, Nielsen J (2013) Inverse identification of wheel–rail contact forces based on observation of wheel disc strains: an evaluation of three numerical algorithms. Veh Syst Dyn Int J Veh Mech Mobility 51(1):74–90
10. Eliseev K (2016) Modern methods of contact forces between wheelset and rails determining. In: Evgrafov A (ed) Advances in mechanical engineering. Lecture notes in mechanical engineering. Springer International Publishing, Switzerland, pp 57–66
11. Saidova A, Orlova A (2013) Development of mathematical models of cars carts 18-9810 and 18-9855 for the study of wheel wear. Bull Dnepropetrovsk Natl Univ Railway Transp named after acad. V. Lazaryana–2013 2(44):118–123 (rus)

Some Characteristics of Linear Acceleration Reproduction with Flexible Harmonical Component

Alexander N. Evgrafov, Vladimir I. Karazin, Denis P. Kozlikin
and Igor O. Khlebosolov

Abstract This article shows the possibility of increasing the range of experimental actions on a vibro-rotary machine (vibrafuge) by exploiting the properties of resonance phenomena in the structure and considers the feasibility of using self-resonance operating conditions.

Keywords Centrifugal machine · Vibrafuge · Electrodynamic shaker
Combined action · Resonance system · Self-resonance

Introduction

This study is dedicated to solving the problems of permanent and variable force action testing of devices under laboratory conditions. Centrifuges have a significant advantage over linear acceleration stands. On a rotary system, the time of rotation required to reach normal acceleration is not limited. Therefore, tests can run for as long as is needed to obtain the intended result. Harmonic action on normal linear acceleration can be achieved using an additional device positioned on a rotor of the centrifuge and equipped with its own independent drive. Publications [1–7] provide a rather extensive overview of the devices suitable for said action. However, there are no known cases of successful implementation of any engineering solution that would provide a significantly wide range of variation of experimental action.

A. N. Evgrafov (✉) · V. I. Karazin · D. P. Kozlikin · I. O. Khlebosolov
Peter the Great St. Petersburg Polytechnic University, Saint Petersburg, Russia
e-mail: a.evgrafov@spbstu.ru

V. I. Karazin
e-mail: visv05@mail.ru

D. P. Kozlikin
e-mail: kozlikindenis@gmail.com

I. O. Khlebosolov
e-mail: khlebosolov@mail.ru

© Springer International Publishing AG 2018
A. N. Evgrafov (ed.), *Advances in Mechanical Engineering*, Lecture Notes
in Mechanical Engineering, https://doi.org/10.1007/978-3-319-72929-9_9

Task

One of the most promising engineering solutions for simulating the above combined action is the use of a resonance system. The main benefit of this method is its low energy consumption and the capability of extending the range of amplitude-frequency action without additional costs.

Let us assess the feasibility of tests for the three parameter groups specified in the table.

Linear overload is created when the rotor of the centrifuge rotates with an angular speed ω and the mass center of the test item is positioned at a distance R from the rotation axis:

$$w_L = R\omega^2. \tag{1}$$

As previously stated, it is rather easy to achieve linear overload on centrifugal machines (centrifuges). As can be seen from Formula (1), it is enough to select distance R and speed ω. This issue has been examined in a great number of publications [8, 9].

From this moment on, we shall focus on obtaining the variable component of experimental action. Vibration displacement r, vibration acceleration w_V and rotational frequency v share a known dependence:

$$r = \frac{w_V}{v^2} = \frac{w_V}{4\pi^2 f^2}. \tag{2}$$

By substituting values from Table 1 into Formula (2), it can be seen that maximum vibration acceleration values at low frequencies correspond to high vibration displacement values, while higher frequencies require vibration acceleration w_V, which is significantly higher than its maximum value ($w_V \gg w_{V\max}$), to achieve $r_{\max}$.

Therefore, two ranges should be considered: the first range with the focus on $r_{\max}$, the second, on $w_{V\max}$. The boundaries of these areas correspond to the following frequencies:

Table 1 Parameter groups

Parameter group	Dimension	1	2	3	Symbols
Mass	kg	20	100	500	m
Linear overload	m/s^2	2000	500	250	w_L
Harmonic vibration acceleration	m/s^2	500	300	150	w_V
Vibration displacement	mm	12.5	12.5	12.5	r
Shaker expulsive force	kN	13.2			P
Specified frequency range	Hz	10–2000			f

$$v_B = \sqrt{\frac{w_{V\max}}{r_{\max}}}. \tag{3}$$

By substituting values from Table 1, we get

$$v_{IB} = \sqrt{\frac{500}{12.5 \times 10^{-3}}} = 200 \ 1/s, \quad f_{IB} = 31.8 \text{ Hz},$$

$$v_{IIB} = \sqrt{\frac{300}{12.5 \times 10^{-3}}} = 154.9 \ 1/s, \quad f_{IIB} = 24.7 \text{ Hz},$$

$$v_{IIIB} = \sqrt{\frac{150}{12.5 \times 10^{-3}}} = 109.5 \ 1/s, \quad f_{IIIB} = 17.4 \text{ Hz}.$$

Let us use an electrodynamic shaker in order to simulate the entire specified frequency range. Expulsive force P is calculated according to the following formula:

$$P = F \sin(vt + \varphi),$$

where $F = mw_V$ is the peak value of the vibration machine's expulsive force and φ is the phase shift.

$$A = \pi F r \sin(\alpha),$$

where A is the work done during the full vibration cycle and α is the phase shift. With $\alpha = \pi/2$, $A = \pi F r$, average power is

$$N = \frac{1}{2} v F r,$$

Let us evaluate the power change as the vibration frequency increases. In the area of the first range corresponding to the low-frequency area ($r_{\max} = const.$), according to Formula (3),

$$\frac{N_1^{LOW}}{N_2^{LOW}} = \frac{v_1 m \left[r_{\max} v_1^2 \right] r_{\max}}{v_2 m \left[r_{\max} v_2^2 \right] r_{\max}} = \frac{v_1^3}{v_2^3}.$$

In the area of the second range ($w_{V\max} = const.$),

$$\frac{N_1^{HIGH}}{N_2^{HIGH}} = \frac{v_1 m w_{V\max} \frac{w_{V\max}}{v_1^2}}{v_2 m w_{V\max} \frac{w_{V\max}}{v_2^2}} = \frac{v_2}{v_1}.$$

Thus, with work done in the low-frequency area (first range), the power grows with an increase of v, while in the high-frequency area, it drops. Therefore, it reaches its maximum value at the boundary:

$$N_{max} = \frac{1}{2} v_B Fr_{max} = \frac{1}{2} \frac{mw^2_{V\max}}{v_B}. \tag{4}$$

By substituting parameters from Table 1, it is possible to obtain the maximum values of the powers corresponding to each of the previously specified groups:

$$N_{1max} = 12.5 \text{ kW},$$
$$N_{2max} = 29.1 \text{ kW},$$
$$N_{3max} = 36.3 \text{ kW}.$$

The maximum values of the expulsive forces of shakers will be as follows:

$$P_{1max} = 10 \text{ kN},$$
$$P_{2max} = 30 \text{ kN},$$
$$P_{3max} = 75 \text{ kN}.$$

Comparison with the initially specified expulsive force value shows that it is not sufficient for the purposes of the problem in question.

Usage of the Vibrance Property for Problem Solution

Let us explore the possibility of using a resonance system, in which vibration amplitude is known to spike at a specified disturbance force, and the specified vibration amplitude is provided by minimum force action from the vibration generator. It is also known that at resonance, elastic and inertial forces are interbalanced, while the power of the vibration generator is used to overcome dissipative forces [10].

Practical implementation of resonance conditions is hindered by the need for a corresponding resonance configuration, which presents some constructive difficulties. This involves adjusting the natural frequency of the system, which is determined by the values of mass (m) and rigidity (c). The matter is further complicated by the fact that the vibration system itself is usually non-linear, which increases the risk of unstable and physically unrealizable conditions. There are two known variants of resonance curve (Fig. 1), one with a rigid (a) and one with a soft (b) non-linearity.

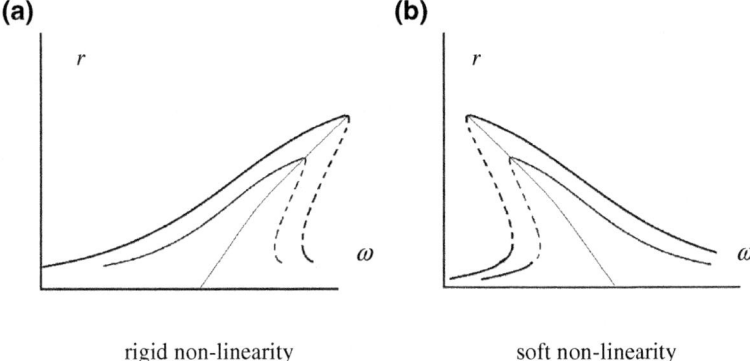

rigid non-linearity   soft non-linearity

Fig. 1 Resonance curves

The type of vibration process depends on the direction of changes in disturbance frequency. Vibration amplitude may also change abruptly.

As a first approximation, let us examine a suspension on three plate springs (Fig. 2a).

A fragment of said suspension, i.e., the calculation model, is shown in Fig. 2b. If the section of the console beam suspension remains unchanged and its mass is negligible compared to the mass of the table and the test item (m), then we get

$$y = \frac{F}{6EJ}\left(x^3 - 3lx^2 + 2l^3\right),$$

where F is the power acting on the suspension, E is the Young's modulus, J is the moment of inertia of the suspension section, l is the distance from the plane of console fixing to the table's center of mass, x is the running coordinate along the console, and y is the console deflection.

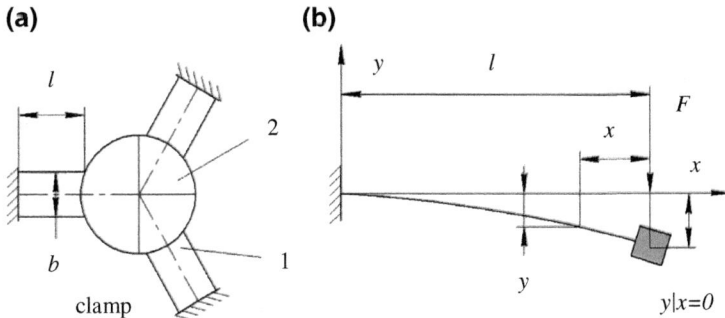

Fig. 2 Calculation model

If $x = 0$, for a weightless beam,

$$v_0 = \sqrt{\frac{c}{m}}, \quad v_0^2 = \frac{c}{m} = \frac{G/y_{st}}{G/g} = \frac{g}{y_{st}}, \quad y_{st} = \frac{mgl^3}{3EJ},$$

where c is the console rigidity toward deformation, G is the gravity force of mass m, y_{st} is the static deflection of the console under the action of the gravity force G, and g is the gravitational acceleration.

Given that $v = 2\pi f$ and $\Delta = r = \frac{w}{v^2}$, for $w = 500$ m/s^2, for example, within the frequency range $f_1 = 1000$ Hz and $f_2 = 100$ Hz, we get

$$\left(v_0^2\right)_1 = 4 \times 10^7 \ 1/s^2, \quad \Delta_1 = r_1 = 1.28 \times 10^{-5} \ m,$$
$$\left(v_0^2\right)_2 = 4 \times 10^5 \ 1/s^2, \quad \Delta_2 = r_2 = 1.28 \times 10^{-3} \ m.$$

If the natural frequency of the system is changed by adjusting the length of the suspension, then

$$c = \frac{3EJ}{l^3}, \quad \frac{\left(v_0^2\right)_1}{\left(v_0^2\right)_2} = \frac{l_2^3}{l_1^3}, \quad \frac{l_2}{l_1} = \sqrt[3]{100} \approx 4.65.$$

For a wider frequency range, for example, $f_1 = 1000$ Hz and $f_3 = 10$ Hz,

$$\frac{f_1}{f_3} = \frac{1000}{10} = 100, \quad \frac{c_{max}}{c_{min}} = 10^4,$$

in which case

$$\frac{l_3}{l_1} = 10^{4/3} \approx 21.5.$$

Changing the natural frequency by adjusting the mass is ineffective, since as mass changes by a factor of two, natural frequency changes by a factor of $\sqrt{2} \approx 1.4$.

Were we to examine a trapezoidal suspension (see Fig. 3) with ratio $b_1 = 8b_2$, we would get, for deflection,

$$y = \frac{12Fl^3}{3Ebh^3},$$

where b is the suspension width and h is its thickness (not shown in the figure).

Fig. 3 Calculation model

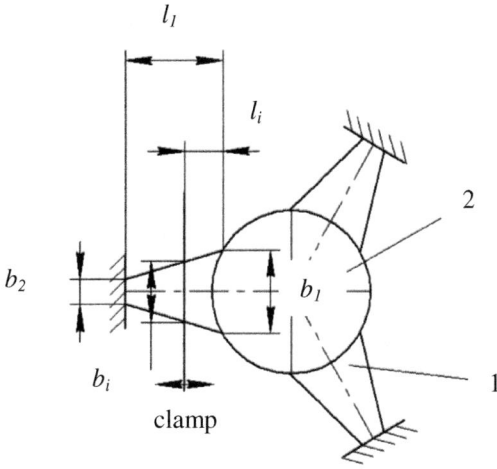

clamp

If $f_1 = 1000$ Hz,　$y_1 = \dfrac{12Fl_1^3}{3Eb_1h^3} = r_1 = 1.25 \times 10^{-5}$ m,

If $f_2 = 100$ Hz,　$y_2 = \dfrac{12Fl_2^3}{3Eb_2h^3} = r_2 = 1.25 \times 10^{-3}$ m,

$$\frac{l_2}{l_1} = \sqrt[3]{\frac{(1000/100)^2}{b_1/b}} \approx 2.32.$$

For range $f_1 = 1000$ Hz and $f_3 = 10$ Hz with the same ratio $b_1 = 8b_2$

$$\frac{l_3}{l_1} \approx 10.8.$$

As an alternative, it is possible to use self-resonance conditions for the vibration system as the most effective way of solving the problem in question [11].

Self-resonance is resonance under the action of the force produced by the movement of the machine's vibration system itself. In this case, periodic self-vibrations are excited through positive feedback that creates a disturbance force depending on the vibration parameters of the actuating element. With specific negotiation of the parameters of the vibration system and the positive feedback chain, the self-resonance system enables self-excited vibrations, automatic maintenance of the resonance state under widely changing load during the technical process. Based on this principle, simple and inexpensive systems are developed, which allow for excitation and stabilization of the most effective resonance conditions of certain vibration machines, such as ultrasonic devices [12, 13] and machines with an unbalance vibration exciter actuated by synchronous alternating current motors [14, 15].

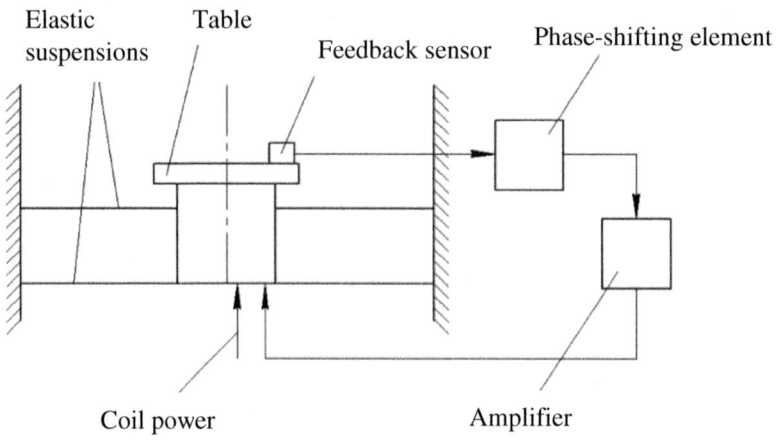

Fig. 4 Self-resonance scheme excitation

Figure 4 shows a diagram of such a system.

The feedback sensor is a vibration sensor that records vibrations and the signal of which is used to generate a sweep signal to the vibrator coil.

The platform coil vibration frequency (parameter ψ) can be adjusted by shifting the signal phase in the feedback chain. A signal phase shift in the feedback chain causes reorganization of the coil vibration conditions (Fig. 5).

Thus, it is possible to adjust the vibration frequency of the coil of the mechanical system by shifting the phase of the sweep signal. Paper [16] claims that during the design of a resonance machine, effective excitation of vibrations is ensured through non-linear transformation of the signal proportional to coil movement. As previously mentioned, under appropriate conditions, the self-resonance system maintains the resonance state of the vibrator under significant load changes. With this in mind, let us go back to the issues of vibrator power and energy dissipation.

The calculation model is presented in the form of a harmonic oscillator [17] with a single degree of freedom and the appropriate selection of mass (m), rigidity (c) and dissipation (b). The vibration equation takes the following form:

$$m\ddot{x} + 2b\dot{x} + cx = F\,\sin(vt),$$

where F is the amplitude of the disturbance force. At resonance, this equation is solved according to the following function:

$$x = r\sin\left(vt - \frac{\pi}{2}\right),$$
$$r = \frac{F}{2bvm}. \tag{5}$$

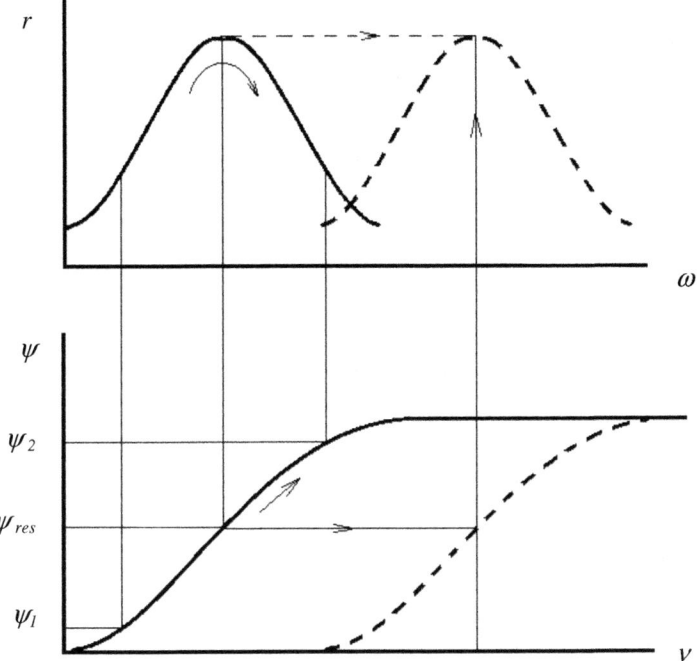

Fig. 5 Scheme of work in self-resonance

Let us determine the power consumed by the vibrator during the vibration period:

$$N = \frac{1}{2\pi v} \int\limits_{0}^{2\pi/v} \dot{x}F \, \sin(vt)dt = \frac{rFv}{2}.$$

By substituting the expression for F from (5), we get

$$N = r^2 b v^2 m.$$

For vibration acceleration 500 m/s^2, $r = \frac{500}{v^2}$ and final power,

$$N = 4.9 \times 10^4 \frac{b}{f^2}, \; W. \tag{6}$$

What makes Formula (6) difficult is the need to determine the dissipation coefficient b, which depends on the frequency of the disturbing force F. If b were constant, it could be concluded that creating vibration with frequency 2000 Hz requires the vibrator power to be 40,000 times lower than that for vibration with a frequency of 10 Hz. Energy dissipation obviously grows as frequency increases. However, the dissipation coefficient is unlikely to increase more than 40, 000 times while passing from 10 to 2000 Hz. Therefore, it can be assumed that vibrator power sufficient for simulation of 10 Hz should be enough to create vibrations with a frequency of 2000 Hz. Some portion of the useful vibrator energy is spent on overcoming the internal non-elastic resistance in the metal. Within the framework of the linear elasticity theory, vibrations of a strip with thickness h under flat deformation conditions are accompanied by the so-called layer movement. Their significance is determined by the dimensionless frequency Ω, which should be lower than 1 ($\Omega < 1$).

$$\Omega = \frac{2}{\pi} f h \sqrt{\frac{v}{E}},$$

If $f = 2000$ Hz, $\Omega = 1.58$ h.

It is possible to determine the value of thickness h, at which layer movements will be significant for energy calculation:

$$h > \frac{1}{1.58} = 0.63 \text{ m}.$$

Conclusion

Based on the above arguments, it can be concluded that in order to obtain the specified experimental actions across the entire frequency range, the method of resonance excitation of vibrations can be used. It makes it possible to significantly reduce the system's energy consumption and simulate experimental actions that have previously been impossible.

References

1. Karazin VI, Kozlikin DP, Sloushch AV, Khlebosolov IO (2007) Dynamic model of vibratory stand. In: Theory of mechanisms and machines, vol 5, no 1 (9). Publishing House of the Polytechnic University Press, St. Petersburg, pp 38–44
2. Karazin VI, Kozlikin DP, Khlebosolov IO (2006) Dynamic stands for vibro-rotary tests, no 3 (45). Scientific and technical reports of SPbGPU. Publishing House of SPbGPU, St. Petersburg, pp 44–49
3. Karazin VI, Kozlikin DP, Khlebosolov IO (2007) On balancing the inertial forces in vibrotsentrifugal. In: Theory of mechanisms and machines, vol 5, no 2 (10). Publishing House of the Polytechnic University Press, St. Petersburg, pp 63–71. Periodic Scientific and Methodical Journal
4. Ksenofontov VI, Nikolaev VN, Chernokrylov SYu (1992) Multifunctional dynamic stand. In: Testing and control stands. Leningrad. LGTU, pp 29–32

5. Rodgers JD, Cericola F, Doggett JW, Young ML (1986) Vibrafuge: combined vibration and centrifuge testing. In: Shok and vibration symposium. SAND89—1659C
6. Doggett J, Cericola F (1989) Vibrafuge—a combined environment testing facility vibration testing on a centrifuge. SAE Technical Paper 892368. https://doi.org/10.4271/892368
7. Jepsen R, Romero E (2005) Testing in a combined vibration and acceleration environment. IMAC XXIII, Orlando, FL
8. Evgrafov AN, Karazin VI, Smirnov GA (1999) Rotary stands for motion variables simulation. In: Scientific and technical reports of SPbGTU, no 3 (17). St. Petersburg, pp 89–94
9. Evgrafov AN, Karazin VI, Khlebosolov IO (2003) Playing motion parameters on rotary stands. In: Theory of mechanisms and machines, vol 1, no 1. Publishing House of the Polytechnic University Press, St. Petersburg, pp 92–96
10. Evgrafov AN, Karazin VI, Kozlikin DP, Khlebosolov IO (2017) Centrifuges for variable accelerations generation. In: International review of mechanical engineering (IREME), vol 11, no 5, pp 280–285
11. Antipov VI, Astashev VK (2004) About the principles of energy-saving machines creation. J Mach Manuf Reliab 4:3–8
12. Astashev VK, Babitsky VI, Vulfson II e.a. (1988) Handbook: dynamics of machines and machine control. In: Kreinin GV Mashinostroenie, Moscow, 329 p
13. Astashev VK, Babitsky VI, Sokolov IY (1990) Autoresonant vibration excitation by synchronous motor. J Mach Manuf Reliab (4):41–46
14. Astashev V, Hertz M (1976) The excitation and stabilisation of resonant vibration in ultrasonic rod systems. Acoust J 22(2):192–200
15. Astashev V, Babitsky V (1982) Methods of ultrasonic machine efficiency increase. In: Stanki I Instrumenti (Machine-tools and tools) no 3, pp 25–27
16. Astashev VK, Babitsky VI (2007) Ultrasonic processes and machines: dynamics, control and applications. Springer, Berlin, 330 p
17. Panovko YaG (1980) Introduction to mechanical vibrations. Science, Moscow, 272 p

Self-braking of Planar Linkage Mechanisms

Alexander N. Evgrafov and Gennady N. Petrov

Abstract The processes of self-braking of Assur group linkage mechanisms are considered in this work. The conditions are determined under which the processes of self-braking and de-braking occur.

Keywords Self-braking · De-braking · Assur group · Structural group
Ideal kinematic pair · Friction

Introduction

It is important to take into account frictional forces in linkage mechanisms if we approach the special positions of structural groups for which the effect of self-braking can occur. Let us determine the conditions that must be fulfilled in order to ensure that self-braking and de-braking do not occur, using the example of the simplest structural groups.

Task

The method of compilation of equilibrium equations and equations of kinematic analysis of planar linkage mechanisms is used in the work [1]. The method is based on the opening of closed structural groups and their reduction to a tree structure. It is shown that the value of the D determinant that is common for such systems can serve as a criterion for the quality of the position of each structural group. Let us consider the mechanism that is taken as an example in this work (Fig. 1a).

A. N. Evgrafov (✉) · G. N. Petrov
Peter the Great St. Petersburg Polytechnic University, Saint Petersburg, Russia
e-mail: a.evgrafov@spbstu.ru

G. N. Petrov
e-mail: gnpet@mail.ru

© Springer International Publishing AG 2018
A. N. Evgrafov (ed.), *Advances in Mechanical Engineering*, Lecture Notes
in Mechanical Engineering, https://doi.org/10.1007/978-3-319-72929-9_10

83

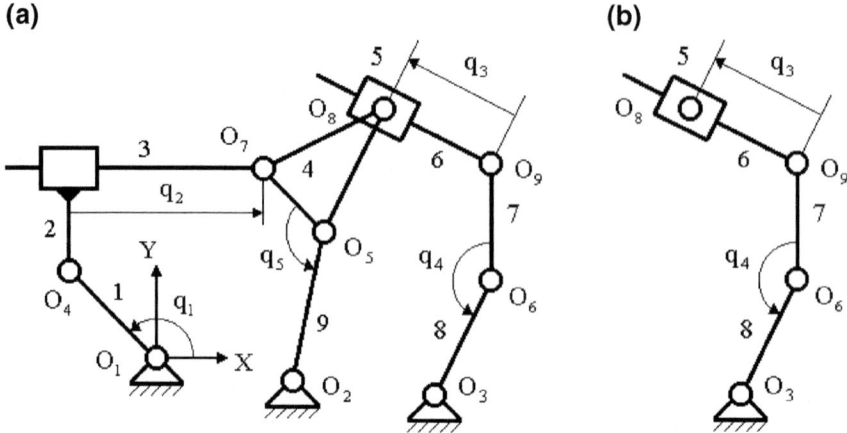

Fig. 1 Example of planar linkage mechanism (**a**) kinematic diagram, (**b**) last structural group

The condition is obtained for the last structural group of the mechanism (Fig. 1b), under which self-braking is possible, if substantial frictional forces arise in the passive joint O_9:

$$\rho > D/O_3O_8 = h, \tag{1}$$

where

ρ is the radius of the friction circle,
D is the double area of the triangle $O_3O_8O_9$, and
h is the height of this triangle drawn from the point O_9.

Figure 2a shows the general view of a single-loop structural group with three passive joints. It consists of kinematic groups A (links $A', \ldots, A''$) and B (links $B', \ldots, B''$). The internal kinematic pairs (KP) of these groups must be active. The links A' and B' are connected by external passive joints 1 and 2 with links C and E of the mechanism, respectively. The internal passive joint 3 connects links A'' and B''. For example, for the structural group (Fig. 1b), the kinematic group A consists of links 5 and 6 with an active prismatic kinematic pair q_3, and the group B consists of links 7 and 8 with an active revolute KP q_4. In the special case, if the structural group does not have internal inputs, then it comes to the Assur group (two links with three passive joints).

Let us assume that we know the reactions in the KP without taking friction forces into account: $\overline{\mathbf{R}}_1^0$, $\overline{\mathbf{R}}_2^0$ are the reactions of links C and E to links A' and B', and $\overline{\mathbf{R}}_{12}^0$ is the reaction of link A'' to link B'' $\left(\overline{\mathbf{R}}_{12}^0 = -\overline{\mathbf{R}}_{21}^0\right)$.

The task of determining these vectors is reduced to solving a linear system of algebraic equations [2]. The total reactions (taking into account friction forces) will be found in the form

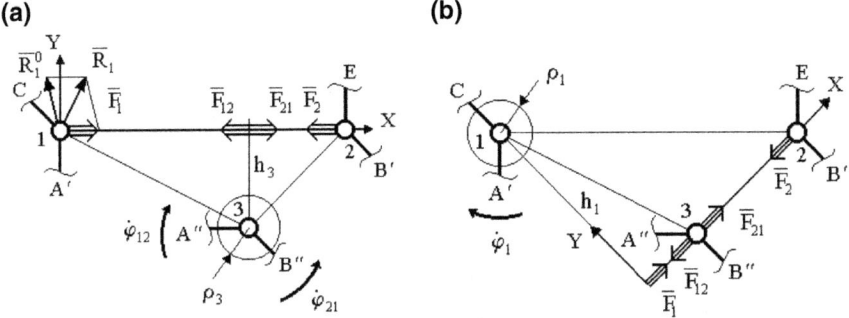

Fig. 2 Determination additional reactions (**a**) friction in joint 3, (**b**) friction in joint 1

$$\overline{\mathbf{R}}_1^1 = \overline{\mathbf{R}}_1^0 + \overline{\mathbf{F}}_1; \quad \overline{\mathbf{R}}_2^1 = \overline{\mathbf{R}}_2^0 + \overline{\mathbf{F}}_2; \quad \overline{\mathbf{R}}_{12}^1 = \overline{\mathbf{R}}_{12}^0 + \overline{\mathbf{F}}_{12}.$$

For the balance of the forces acting on the groups of links A and B to remain unchanged, the vectors $\overline{\mathbf{F}}_1, \overline{\mathbf{F}}_2, \overline{\mathbf{F}}_{12}, \overline{\mathbf{F}}_{21}$ must be put on the same straight line and be in balance with one another:

$$\overline{\mathbf{F}}_1 = \overline{\mathbf{F}}_{12} = -\overline{\mathbf{F}}_{21} = -\overline{\mathbf{F}}_2 = \overline{\mathbf{F}}. \tag{2}$$

Friction in Passive Rotational Pairs

Let us define additional reactions if we take friction into account only in the internal passive joint (Fig. 2a). We introduce a local coordinate system XY so that the axis X passes through two passive joints. It is obvious that the line of action of the additional forces must coincide with the axis X. The expression for the moment in the inner joint is as follows:

$$|Fh_3| = \varepsilon_{21} F |h_3| = \rho_3 \left| \overline{\mathbf{R}}_{12}^0 + \overline{\mathbf{F}} \right|, \tag{3}$$

where F is the projection of the additional force $\overline{\mathbf{F}}$ on the axis X, $\varepsilon_{21} = \text{sign}(\dot{\varphi}_{21}) = -\text{sign}(\dot{\varphi}_{12})$ is the sign of the projection of the additional force, which is determined by the direction of the relative angular velocity of the links A'' and B'' (at $\varepsilon_{21} = 1$, the direction of the additional forces coincides with the direction in Fig. 2a),

h_3 is the distance from joint 3 to the line of action of the additional forces, and ρ_3 is the radius of the friction circle.

From (3), we obtain the following values for the force F:

$$F = \frac{R_{12X} \pm \sqrt{R_{12X}^2 + \frac{h_3^2 - \rho_3^2}{\rho_3^2} R_{12}^2}}{\frac{h_3^2 - \rho_3^2}{\rho_3^2}} = \frac{R_{12X} \pm \sqrt{R_{12X}^2 + (k_{h3}^2 - 1)R_{12}^2}}{k_{h3}^2 - 1}$$

where R_{12X}, R_{12Y} are the projections of the vector $\overline{\mathbf{R}}_{12}^0$ onto the local axes X and Y and $k_{h3} = \frac{|h_3|}{\rho_3}$ is an assembly assurance factor.

If an assembly assurance factor $k_{h3} > 1$, we obtain a single solution for the F (the sign '+' is chosen if $\varepsilon_{21} = 1$, the sign '−' is chosen if $\varepsilon_{21} = -1$):

$$F = \frac{R_{12X} + \varepsilon_{21} \sqrt{R_{12X}^2 + (k_{h3}^2 - 1)R_{12}^2}}{k_{h3}^2 - 1},$$

where $R_{12} = \left| \overline{\mathbf{R}}_{12}^0 \right|$.

Otherwise $(k_{h3} < 1)$, we get two solutions (de-braking) or none (self-braking). These conclusions coincide with the condition (1).

Let us draw the analogous expression for the force F, taking into account the friction in the passive joint 1 (Fig. 2b):

$$F = \frac{R_{1X} + \sqrt{R_{1X}^2 + (k_{h1}^2 - 1)R_1^2}}{k_{h1}^2 - 1},$$

where $k_{h1} = \frac{|h_1|}{\rho_1}$, $R_1 = \left| \overline{\mathbf{R}}_1^0 \right|$.

Friction in Passive Sliding Pairs

Let us now consider a structural group with a passive prismatic kinematic pair (Fig. 3a).

Let us introduce an additional local coordinate system $X_3 Y_3$ associated with the link B''. The axis X_3 is directed along the line of relative movement of the links A'' and B''. The coordinate φ_3 specifies the angle between the axes X and X_3. Let us determine the value of the frictional force for the model of the prismatic KP in the case of a two-point contact (Fig. 3b):

$$|F_{\text{fr}}| = f \left(\left| \frac{R_3}{2} + \frac{M_3}{\ell_3} \right| + \left| \frac{R_3}{2} - \frac{M_3}{\ell_3} \right| \right) = f \left(\frac{1 + \varepsilon_p}{2} |R_3| + \frac{1 - \varepsilon_p}{\ell_3} |M_3| \right), \quad (4)$$

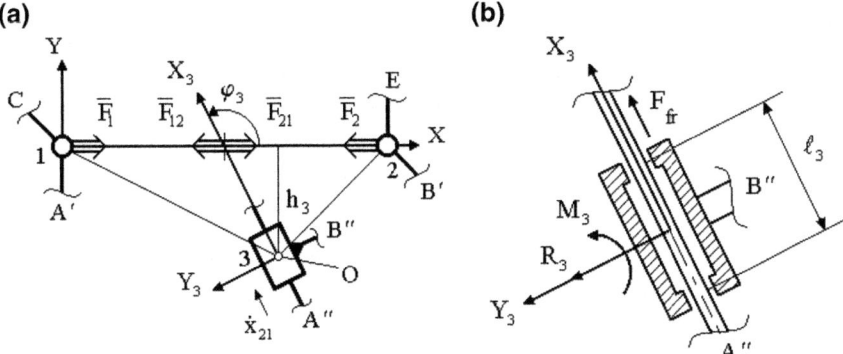

Fig. 3 Structural group with a passive prismatic kinematic pair (**a**) determination additional reactions, (**b**) the model of the prismatic kinematic pair in the case of a two-point contact

where $f = \text{tg}\,\alpha$ is the friction coefficient of sliding,

α is the angle of friction,

ℓ_3 is the length of the stone,

and R_3, M_3 are the normal reaction and the moment in the prismatic pair, for the point O,

$$\varepsilon_p = \text{sign}\left(\left|\frac{R_3}{2}\right| - \left|\frac{M_3}{\ell_3}\right|\right).$$

Let us introduce additional forces directed along the axis X and satisfying condition (2). F_{fr}, R_3, M_3 are defined, taking into account the strength $\overline{F}_{12}$:

$$|F_{\text{fr}}| = |F \cos \varphi_3|; \quad |R_3| = |R_{12} + F \sin \varphi_3|; \quad |M_3| = |M_{12} - Fh_3|, \quad (5)$$

where R_{12} is the projection of the reaction of the link A'' to the link B'' onto axis Y_3 (if we do not take into account the frictional forces), and h_3 is the coordinate of the point O along the axis Y.

We put expressions (5) into (4):

$$|F \cos \varphi_3| = \varepsilon_M \varepsilon_{12} F \cos \varphi_3 = f\left(\frac{1 + \varepsilon_p}{2}|R_{12} + F \sin \varphi_3| + \frac{1 - \varepsilon_p}{\ell_3}|M_{12} - Fh_3|\right),$$

$$(6)$$

where $\varepsilon_{12} = \text{sign}\,\dot{x}_{12}$ is the sign of the projection of force F on the axis X_3, which is determined by the direction of the relative linear speed of the links A'' and B'', and $\varepsilon_M = \text{sign}(\cos \varphi_3)$ is «the assembly sign» of the structural group.

Let us consider two possible cases.

1. One-sided contact in the prismatic pair ($\varepsilon_p = 1$). The additional force is determined from (6):

$$F = \frac{\varepsilon_N R_{12} f}{\varepsilon_{12}\varepsilon_M \cos \varphi_3 - \varepsilon_N f \sin \varphi_3} = \frac{\varepsilon_{12}\varepsilon_M \varepsilon_N R_{12}\left(\frac{1}{f}\cos \varphi_3 + \varepsilon_{12}\varepsilon_M \varepsilon_N \sin \varphi_3\right)}{\frac{1}{f^2}\cos^2 \varphi_3 - \sin^2 \varphi_3}, \quad (7)$$

where $\varepsilon_N = \mathrm{sign}(R_{12} + F \sin \varphi_3)$.

Likewise with the revolute pair, we introduce the concept of the assembly assurance factor:

$$k_{h3}^2 - 1 = \frac{1}{f^2}\cos^2 \varphi_3 - \sin^2 \varphi_3 \Rightarrow k_{h3} = \left|\frac{\cos \varphi_3}{\sin \alpha}\right|. \quad (8)$$

We put (8) into (7):

$$F = \frac{\varepsilon_{12}\varepsilon_M \varepsilon_N R_{12} \cos(\varphi_3 - \varepsilon_{12}\varepsilon_M \varepsilon_N \alpha)}{\sin \alpha \left(k_{h3}^2 - 1\right)}. \quad (9)$$

2. Two-way contact in the prismatic pair $\left(\varepsilon_p = -1\right)$. From (6), we obtain the value of the additional force:

$$F = \frac{2M_{12} f \varepsilon_N}{\ell_3\left(\varepsilon_{12}\varepsilon_M \cos \varphi_3 + \varepsilon_N f \frac{2h_3}{\ell_3}\right)} = \varepsilon_{12}\varepsilon_M \varepsilon_N \frac{2M_{12}}{\ell_3}\frac{\left(\frac{1}{f}\cos \varphi_3 - \varepsilon_{12}\varepsilon_M \varepsilon_N \frac{2h_3}{\ell_3}\right)}{\left(\frac{1}{f^2}\cos^2 \varphi_3 - \frac{4h_3^2}{\ell_3^2}\right)}, \quad (10)$$

where $\varepsilon_N = \mathrm{sign}(M_{12} - F h_3)$.

The assembly assurance factor is as follows:

$$k_{h3}^2 - 1 = \frac{1}{f^2}\cos^2 \varphi_3 - 4\frac{h_3^2}{\ell_3^2} \Rightarrow k_{h3} = \sqrt{\frac{f^2 + \cos^2 \varphi_3}{f^2} - 4\frac{h_3^2}{\ell_3^2}}. \quad (11)$$

We put (11) into (10):

$$F = \varepsilon_{12}\varepsilon_M \varepsilon_N \frac{2M_{12}}{\ell_3}\frac{\left(\frac{1}{f}\cos \varphi_3 - \varepsilon_{12}\varepsilon_M \varepsilon_N \frac{2h_3}{\ell_3}\right)}{\left(k_{h3}^2 - 1\right)}. \quad (12)$$

Likewise in the case of a revolute pair, it follows from (9) and (12) that the condition $k_{h3} > 1$ is sufficient for the avoidance of self-braking or de-braking in the work of the structural group.

Friction in All Passive Joints of a Structural Group

We now take into account the frictional forces in all three joints (Fig. 4).

Local coordinate systems $X_0 Y_0$ and XY are introduced so that the axis X_0 passes through two external joints (line LN), and the axis X coincides with the line of action of the additional forces. The shortest distances from the points L, N, M (the axes of the passive joints) to the axis X will be denoted by h_1, h_2, h_3. The origin of the zero coordinate system is chosen at the point L, and the origin of the system XY is chosen at the point O, with the axis Y passing through the point L, and with the value $h_1 > 0$ (in the projection onto the axis Y). The angles φ, φ_1, φ_2 connect the axis X_0, respectively, with the axis X and with the single vectors $\bar{n}_1$ and $\bar{n}_2$, which are directed from points L and N to point M.

Now let us introduce the following notations:

$$\varepsilon_1 = \text{sign}(\dot{\varphi}_1); \quad \varepsilon_2 = \text{sign}(\dot{\varphi}_2); \quad \varepsilon_{12} = \text{sign}(\dot{\varphi}_{12}),$$

where

$\dot{\varphi}_1, \dot{\varphi}_2$	are the angular velocities of the links A' and B' relative to the links C and E;
$\dot{\varphi}_{12}$	is the angular velocity of the link A'' relative to B'';
$\varepsilon_F = -\varepsilon_1$	is the sign of the projection of the additional force on the axis X;
$\varepsilon_{h3} = -\dfrac{\varepsilon_{12}}{\varepsilon_1}$	is the sign of the projection of the vector $\overline{O}\,\overline{M}$ on the axis Y (sign h_3);
and $\varepsilon_{h2} = -\dfrac{\varepsilon_2}{\varepsilon_1}$	is the sign of the projection of the vector $\overline{O}\,\overline{N}$ on the axis Y (sign h_2).

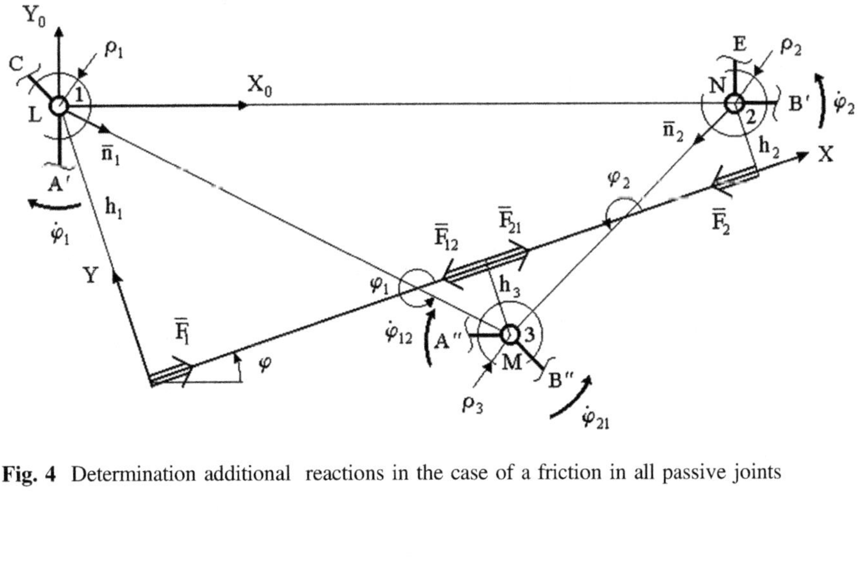

Fig. 4 Determination additional reactions in the case of a friction in all passive joints

The equations for determining the additional forces are the following:

$$\begin{cases} |h_1 F| = -\varepsilon_1 h_1 F = \rho_1 \left| \overline{\mathbf{F}} + \overline{\mathbf{R}}_1^0 \right|; \\ |h_3 F| = \varepsilon_{12} h_3 F = \rho_3 \left| \overline{\mathbf{F}} + \overline{\mathbf{R}}_{12}^0 \right|; \\ |h_2 F| = \varepsilon_2 h_2 F = \rho_2 \left| -\overline{\mathbf{F}} + \overline{\mathbf{R}}_2^0 \right|; \\ \frac{h_3 - h_1}{\ell_1} = \sin(\varphi_1 - \varphi); \\ \frac{h_3 - h_2}{\ell_2} = \sin(\varphi_2 - \varphi), \end{cases} \tag{13}$$

where ρ_1, ρ_2, ρ_3 are the radii of the friction circles of the first, second and third passive joints, and ℓ_1, ℓ_2 are the distances between the passive joints $(\ell_1 = |\overline{\mathbf{L}\,\mathbf{M}}|, \ell_2 = |\overline{\mathbf{M}\,\mathbf{N}}|)$.

The force F, angle φ and distances h_1, h_2, h_3 are unknown values in the system of Eqs. (13). Let us transform this system of equations and get rid of the unknown h_1, h_2, h_3:

$$\begin{aligned} \sin(\varphi_1 - \varphi) &= \frac{\varepsilon_{12}\rho_3 \left| \overline{\mathbf{F}} + \overline{\mathbf{R}}_{12}^0 \right| + \varepsilon_1 \rho_1 \left| \overline{\mathbf{F}} + \overline{\mathbf{R}}_1^0 \right|}{\ell_1 F} = a; \\ \sin(\varphi_2 - \varphi) &= \frac{\varepsilon_{12}\rho_3 \left| \overline{\mathbf{F}} + \overline{\mathbf{R}}_{12}^0 \right| - \varepsilon_2 \rho_2 \left| -\overline{\mathbf{F}} + \overline{\mathbf{R}}_2^0 \right|}{\ell_2 F} = b. \end{aligned} \tag{14}$$

We express $\sin \varphi$ and $\cos \varphi$ through the variables a and b:

$$\cos \varphi = \frac{b \cos \varphi_1 - a \cos \varphi_2}{\sin(\varphi_2 - \varphi_1)};$$

$$\sin \varphi = \frac{b \sin \varphi_1 - a \sin \varphi_2}{\sin(\varphi_2 - \varphi_1)}.$$

Let us square these expressions and sum them up:

$$1 = \frac{a^2 + b^2 - ab \cos(\varphi_2 - \varphi_1)}{\sin^2(\varphi_2 - \varphi_1)} \tag{15}$$

If we take the values of a and b from (14), then the expression (15) can be used to determine the unknown force F.

Let us now consider the movement of the structural group, which assumes that the sum of all forces, including the inertia forces acting on each of the groups of links connected by passive joints, is equal to zero. In this case, we have the following reaction ratio:

$$\overline{R}_1^0 = \overline{R}_{12}^0 = -\overline{R}_{21}^0 = -\overline{R}_2^0 = \overline{R}.$$

From Eqs. (13), we obtain the following expression:

$$k_h = \frac{h_1}{\rho_1} = \frac{|h_2|}{\rho_2} = \frac{|h_3|}{\rho_3} = \frac{|\overline{F} + \overline{R}|}{|\overline{F}|},$$

that is, the assembly assurance factors for all joints are equal. Such an assembly assurance factor is referred to as geometric. From (14), we define the values of a and b:

$$a = -k_h \frac{\varepsilon_{12}\rho_3 + \varepsilon_1\rho_1}{\varepsilon_1 \ell_1}; \quad b = -k_h \frac{\varepsilon_{12}\rho_3 - \varepsilon_2\rho_2}{\varepsilon_1 \ell_2}.$$

From (15), we obtain the following expression:

$$\sin^2(\varphi_2 - \varphi_1) = k_h^2 \left[\left(\frac{\varepsilon_{12}\rho_3 + \varepsilon_1\rho_1}{\ell_1} \right)^2 + \left(\frac{\varepsilon_{12}\rho_3 - \varepsilon_2\rho_2}{\ell_2} \right)^2 \right.$$
$$\left. - \frac{(\varepsilon_{12}\rho_3 + \varepsilon_1\rho_1)(\varepsilon_{12}\rho_3 - \varepsilon_2\rho_2)}{\ell_1 \ell_2} \cos(\varphi_2 - \varphi_1) \right]$$

$$k_h = \frac{|\sin(\varphi_2 - \varphi_1)|}{\sqrt{\left(\frac{\varepsilon_{12}\rho_3 + \varepsilon_1\rho_1}{\ell_1} \right)^2 + \left(\frac{\varepsilon_{12}\rho_3 - \varepsilon_2\rho_2}{\ell_2} \right)^2 - \frac{(\varepsilon_{12}\rho_3 + \varepsilon_1\rho_1)(\varepsilon_{12}\rho_3 - \varepsilon_2\rho_2)}{\ell_1 \ell_2} \cos(\varphi_2 - \varphi_1)}}$$

For the action line of the additional forces (axis X) to pass as shown in Fig. 4, the following conditions must be fulfilled:

$$\varepsilon_{h2} = -\frac{\varepsilon_2}{\varepsilon_1} = -1; \quad \varepsilon_{h3} = -\frac{\varepsilon_{12}}{\varepsilon_1} = 1.$$

If the geometric factor of the assembly stock at the indicated signs ε_{h2}, c_{h3} is less than or equal to 1, it is impossible to select the reactions $\overline{R}_1^0$, $\overline{R}_2^0$, $\overline{R}_{12}^0$ that would provide the avoidance of the unfavorable modes (self-braking or de-braking) in all joints.

In general, when the structural group approaches the described mode of movement, the assembly assurance factors tend to reach the value of one, and a sharp increase in the frictional forces occurs. The structural group passes through the dangerous stage, subjected to an internal impact (the friction forces cause a sudden change in inertia forces). If the change in inertial forces with the increase in the frictional forces does not lead to the end of the dangerous mode, it results in self-braking.

Conclusion

The described algorithm allows us to learn how structural groups of the mechanism approach the positions in which frictional forces significantly increase and the processes of self-braking or de-braking can evolve, which has an unfavorable effect on the performance of the mechanism.

References

1. Kolovsky MZ (1997) On a criterion for the quality of multi-link linkage mechanisms. In: Problems of machine building and machine reliability, no 2, pp 92–98
2. Singer IL, Pollock H (eds) (1992) Fundamentals of friction. Springer, 603 p
3. Mata A, Torras A, Carrillo J, Juanco F, Fernández A, Martínez F (2016) Fundamentals of machine theory and mechanisms. Springer, 409 p
4. Persson BNJ (2000) Sliding friction. Springer, 497 p
5. Mkrtychev OV (2013) Computer simulation with force calculation of planar mechanisms. In: Theory of mechanisms and machines, vol 11, no 1 (21), pp 77–83
6. Ziborov KA (2010) Force analysis of mechanisms using the Mathcad program. In: Ziborov KA, Matsyuk IN, Shlyakhov EM (eds) Theory of mechanisms and machines, vol 8, no 1, pp 83–88
7. Doronin FA (2014) Power analysis of some spatial structures and mechanisms using the MathCad package. In: Theory of mechanisms and machines, vol 12, no 1 (23), pp 59–69
8. Yarullin MG, Khabibullin FF (2017) Theoretical and practical conditions of bennett mechanism workability. In: Evgrafov A (ed) Advances in mechanical engineering. Lecture notes in mechanical engineering. Springer International Publishing, Switzerland, pp 145–153
9. Petrov GN (1993) Algorithm of kinetostatic computer-based calculation of closed linkage mechanisms. In: Problems of machine building and reliability of machines, no 3
10. Evgrafov AN, Petrov GN (2016) Drive selection of the multidirectional mechanism with excess inputs. In: Evgrafov A (ed) Advances in mechanical engineering. Lecture notes in mechanical engineering. Springer International Publishing, Switzerland, pp 31–37
11. Evgrafov AN, Petrov GN (2017) Computer simulation of mechanisms. In: Evgrafov A (ed) Advances in mechanical engineering. Lecture notes in mechanical engineering. Springer International Publishing, Switzerland, pp 45–56
12. Semenov YuA, Semenova NS (2015) Theory of mechanisms and machines in examples and tasks, Part 1: textbook. Publishing house of Polytechnic University, Saint-Petersburg, p 286
13. Evgrafov AN, Petrov GN (2003) Geometric and kinetostatic analysis of planar linkage mechanisms of the second class. In: Theory of mechanisms and machines, no 2, pp 50–63
14. Evgrafov AN, Kolovsky MZ, Petrov GN (2015) Theory of mechanisms and machines: a textbook. Publishing house of Polytechnic University, Saint-Petersburg, p 248
15. Kolovsky MZ, Evgrafov AN, Semenov YuA, Slousch AV (2000) Advanced theory of mechanisms and machines. Springer-Verlag, Berlin, p 394

Waves with the Negative Group Velocity in the Cylindrical Shell, Filled with Compressible Liquid

George V. Filippenko

Abstract The problem of joint oscillations of the infinite thin cylindrical shell filled with acoustical liquid of the Kirchhoff–Love type is considered. Free harmonic vibrations of the system are found. Propagating waves are analyzed. Much attention is given to exploration of waves with negative group velocity in the neighborhood of the bifurcation point of dispersion curves. Dispersion curve asymptotics are used in the neighborhood of the bifurcation point for this case. The ranges of frequencies and wavenumbers where this effect is observed are also estimated. Asymptotics for the regular case and for the case of bifurcation are discussed. Dependence of processes on the relative thickness of the shell and other parameters of the system are viewed. Possible fields of applicability of the effects gained are established.

Keywords Propagation of the waves · Cylindrical shell · Vibrations of the shells Local and integral energy fluxes

Introduction

Cylindrical shells have rich applications in the modern technique. They are important elements of the different acoustic waveguides and pipelines [1–5]. The presence of the liquid making contact with the shell considerably complicates the wave processes [6–20]. The following types of wave process in the system, "shell, filled with the liquid" and "shell, surrounded by liquid", were considered in the papers [6–9, 15–20] and [11, 12] correspondingly.

G. V. Filippenko (✉)
Institute of Mechanical Engineering of RAS, Vasilievsky Ostrov,
Bolshoy Prospect 61, 199178 Saint Petersburg, Russia
e-mail: g.filippenko@spbu.ru

G. V. Filippenko
Saint Petersburg State University, 7-9, Universitetskaya Nab, 199034 Saint Petersburg,
Russia

© Springer International Publishing AG 2018
A. N. Evgrafov (ed.), *Advances in Mechanical Engineering*, Lecture Notes
in Mechanical Engineering, https://doi.org/10.1007/978-3-319-72929-9_11

Besides exploration of kinematic and dynamic characteristics in different shells, types of energy analysis are used, including energy flux analysis, which is associated with the group velocity of the waves [7–22]. This work continues with exploration of the negative group velocity effect and bifurcation of the dispersion curves, which was fulfilled in [9].

Statement of the Problem and Determination of the General Representation for Acoustic and Vibrational Fields

The statement of the problem is similar to that considered in [7]. Let us start by considering an infinite cylindrical shell having radius R, thickness h and filled with an ideal compressible liquid. Its density is ρ_w, and the velocity of sound is equal to c_w. The source of an acoustic field in a wave guide is the vibrations of the cylinder shell, caused by the incident wave propagating from the infinite part ($z = -\infty$) of the shell. The frequency of this incident harmonic wave is equal to ω. All processes in the shell–liquid system are supposed to be harmonic with this frequency. The factor $e^{-i\omega t}$ describing the time-dependence will be omitted.

The acoustic pressure P is described by the Helmholtz equation in the cylindrical system of coordinates (r, z, φ), where the axis z coincides with axis of the cylinder

$$(\Delta + k^2)P(r, z, \varphi) = 0; \quad k = \omega/c_w, \tag{1}$$

$$0 \leq r < R, \quad 0 \leq \varphi \leq 2\pi, \quad -\infty < z < +\infty.$$

Two relations take place on the shell–fluid boundary: kinematic (the impenetrability condition)

$$u_n(z, \varphi) = \frac{1}{\rho_w \omega^2} \frac{\partial P(r, z, \varphi)}{\partial r}\bigg|_{r=R} \tag{2}$$

and dynamic (balance of forces acting on the shell)

$$\frac{\rho c_s^2}{R^2} \mathbf{L_w u} = (0, 0, P)^T. \tag{3}$$

Here, $\mathbf{u} = (u_t, u_z, u_n)^T$ (T means transposition) is the displacement vector of the shell and $\mathbf{L_w}$ is the 3×3 matrix differential operator of the cylindrical shell of the Kirchhoff–Love type

$$\mathbf{L_w} = w^2 \mathbf{I} + \mathbf{L},$$

where $\mathbf{I}$ is the unit matrix operator and $\mathbf{L}$ is the 3×3 matrix differential operator with the following elements (see [7]):

$$L_{11} = \alpha_1 [\partial_\varphi^2 + v_- \partial_z^2], \quad L_{12} = v_+ \partial_z \partial_\varphi,$$

$$L_{12} = v_+ \partial_z \partial_\varphi, \quad L_{13} = \partial_\varphi (1 + 2\alpha^2 [1 - \partial_\varphi^2 - \partial_z^2]),$$

$$L_{22} = v_- \partial_\varphi^2 + \partial_z^2, \quad L_{23} = v \partial_z,$$

$$L_{33} = \alpha^2 (2\partial_\varphi^2 - 1 + 2v\partial_z^2 - [\partial_\varphi^2 + \partial_z^2]^2) - 1,$$

$$L_{21} = L_{12}, \quad L_{31} = -L_{13}, \quad L_{32} = -L_{23},$$

where $\partial_z := R\frac{\partial}{\partial z}$, $\alpha_1 = 1 + 4\alpha^2$, $v_\pm = (1 \pm v)/2$.

The properties of the cylinder's material are characterized by E, v and ρ_s—Young's module, the Poisson coefficient and volumetric density, respectively. We define the surface density of the shell $\rho = \rho_s h$ and the velocity of median surface deformation waves of the cylindrical shell as $c_s = \sqrt{E/((1 - v^2)\rho_s)}$.

The following dimensionless parameters are used: $\alpha = h/(R\sqrt{12})$ (the relative thickness of the cylindrical shell) and $w = \omega R/c_s$ (the dimensionless frequency). An important role in shell–fluid interaction is played by the dimensionless velocity $c = c_s/c_w$ and the dimensionless density $\rho^* = \rho_s/\rho_w$.

It can be noted that the dispersion analysis of a "dry" cylindrical shell of this kind was fulfilled in [4, 5].

Determination of the general representation for acoustic and vibrational fields is similar to that considered in [9]. The new vector $(u_t, u_z, P)^T$ is defined using (2):

$$\begin{pmatrix} u_t \\ u_z \\ u_n \end{pmatrix} = \mathbf{M} \begin{pmatrix} u_t \\ u_z \\ P \end{pmatrix}; \tag{4}$$

$$\mathbf{M} = \begin{pmatrix} 1 & 0 & 0 \\ 0 & 1 & 0 \\ 0 & 0 & \frac{1}{\rho_w \omega^2} \frac{\partial}{\partial r}\big|_{r=R} \end{pmatrix}.$$

Thus, Eq. (3) can be rewritten in the form

$$\mathbf{S} \begin{pmatrix} u_t \\ u_z \\ P \end{pmatrix} = \begin{pmatrix} 0 \\ 0 \\ 0 \end{pmatrix} \equiv \mathbf{0}; \tag{5}$$

$$\mathbf{S} = \mathbf{LM} + \mathbf{N}, \quad \mathbf{N} = \frac{w^2}{\rho_s h \omega^2} \begin{pmatrix} 0 & 0 & 0 \\ 0 & 0 & 0 \\ 0 & 0 & 1 \end{pmatrix}.$$

The solution to Eq. (5) is sought in the form

$$
\begin{pmatrix} u_t \\ u_z \\ P \end{pmatrix} = A e^{i\frac{\lambda}{R}z} \begin{pmatrix} \zeta \sin m\varphi \\ \xi \cos m\varphi \\ \gamma J_m\left(r\sqrt{k^2 - (\lambda/R)^2} \right) \cos m\varphi \end{pmatrix}, \tag{6}
$$

where A, ζ, ξ, γ are arbitrary constants, so that $|\zeta|^2 + |\xi|^2 + |\gamma|^2 = 1$; J_m is a Bessel function with index m and λ is the dimensionless wavenumber that we are looking for. It can be noted that if $k < \lambda/R$, then J_m should be replaced with the Bessel function I_m. After substituting (6) into (5), the following algebraic system is obtained:

$$
\mathbf{S}^* \mathbf{x} = \mathbf{0}; \quad \mathbf{x} = (\zeta, \xi, \gamma)^T. \tag{7}
$$

The operator $\mathbf{S}^*$ is the Fourier image of the operator $\mathbf{S}$. The following dispersion equation is obtained from the condition of existence of a non-trivial solution of the system (7)

$$
\det \mathbf{S}^* = 0. \tag{8}
$$

We are looking for real positive solutions to this equation. If the corresponding set of wavenumbers is found, one can solve Eq. (7) and define the previously unknown constants ζ, ξ, γ. After defining the constants, the complete solution to the problem in terms of displacements of the shell $\mathbf{u}(\varphi, z)$ and pressure $P(r, \varphi, z)$ in the liquid is found.

Asymptotic Analysis of the Dispersion Equation, When $\lambda \to 0$

The dispersion equation (8) in the case of a shell without liquid reduces to the form

$$
\det \mathbf{L}_w^* = 0; \quad \mathbf{L}_w^* = w^2 \mathbf{I} + \mathbf{L}^*, \tag{9}
$$

where operators $\mathbf{L}_w^*$, $\mathbf{L}^*$ are the Fourier images of operators $\mathbf{L}_w$, $\mathbf{L}$. This dispersion equation is a bicubic equation of w^2. It has three real non-negative roots. These roots, as functions of λ, determine three dispersion curves

$$
w_i(\lambda) \equiv \sqrt{w_i^2(\lambda^2)} \geq 0, \quad i = 1, 2, 3.
$$

Their starting points (cutoff frequencies) are designated as $w_i \equiv w_i(0), i = 1, 2, 3$; $w_1 < w_2 < w_3$; $w_2 = m\sqrt{v_-}$. For further consideration, it will be useful to use algebraic adjunctions $M_{i,j}(w^2, \lambda^2)$, $i, j = 1, 2, 3$ of the matrix $\mathbf{L}_w$. Under these terms, the equation for determining w_1, w_3 can be written in the form

$$M_{2,2}(w^2, 0) = 0. \tag{10}$$

The dispersion equation (9) is the implicit function of the second degrees of w and λ and, hence, in the regular case (when the existence condition of an inverse function is fulfilled), asymptotic of the dispersion curve when $\lambda \to 0$ has the view

$$w_j^2(\lambda) = w_j^2 + B_j \lambda^2 + O(\lambda^4); \ B_j > 0; \quad j = 1, 2, 3.$$

But due to the specific choice of parameters m, c and α (if m is rather great for the case of the Kirchhoff–Love model [10]), the two-fold root $w = w_2$ of dispersion equation (9) can appear. This root must satisfy the equation

$$M_{2,2}(w_2^2, 0) = 0. \tag{11}$$

In this situation, starting points w_1 and w_2 of the first and second dispersion curves (they correspond to the first two cutoff frequencies) can coincide (in this case, their abscises are equal to $w_2 \equiv m\sqrt{v_-}$). This point will hereinafter be called the "bifurcation point" for brevity, and dispersion curves $w_1(\lambda)$ and $w_2(\lambda)$ will be designated as the left $w_-(\lambda)$ and right $w_+(\lambda)$ dispersion curves, correspondingly. In this case, it is necessary to look for this asymptotic of the dispersion curve in the form

$$w^2(\lambda) = w_2^2 + A\lambda + B\lambda^2 + O(\lambda^3). \tag{12}$$

After substituting this expression in (9), one can obtain the representations of the coefficients A and B (their asymptotic view will be shown below).

The condition of bifurcation (11) can be simplified in the assumption of a great m (and correspondingly, a small $\varepsilon = 1/m$) and a bounded value of parameter $p = m\alpha$. As a result, (11) obtains the view

$$p^2 = v_- + \frac{3v_-(1 + 2v_-)}{(1 - v_-)} \varepsilon^2 + O(\varepsilon^4). \tag{13}$$

Hence, the asymptotic form of coefficients A and B can be obtained in (12):

$$A = \pm 4v_- + O(\varepsilon^2), \quad B = \frac{3v_-}{2} + O(\varepsilon^2). \tag{14}$$

One can see that these coefficients do not depend on the number of the mode in the first approximation. So, dispersion curves (12) for all m are represented by the unique parabola in variables (w^2, λ). This shows that these processes conform to some extent.

In variables (w, λ), asymptotic expansion for w has the view

$$w_{\pm}(\lambda) = w_2 \pm a\lambda + b\lambda^2 + O(\lambda^3), \tag{15}$$

where

$$a = 2\sqrt{v_-}\,\varepsilon + O(\varepsilon^3), \quad b = \frac{3\sqrt{v_-}}{4}\varepsilon + O(\varepsilon^3). \tag{16}$$

Hence, the maximum of the negative dimensionless group velocity $\frac{dw}{d\lambda} = a$ is reached at $\lambda \to 0$ and the group velocity of the waves from dispersion curves w_- and w_+ has the opposite signs. The module of it can be calculated for any m according to the formula

$$\frac{dw}{d\lambda} = \frac{2\sqrt{v_-}}{m} + O\left(\frac{1}{m^3}\right). \tag{17}$$

The ranges of wavenumbers $\delta\lambda$ and frequencies δw where the negative group velocity can be observed are determined by the formulas

$$\delta\lambda = \frac{4}{3} + O\left(\frac{1}{m^2}\right), \quad \delta w = \frac{4\sqrt{v_-}}{3m} + O\left(\frac{1}{m^3}\right). \tag{18}$$

For $v = 0.28$ (steel), these formulas are transformed into the view

$$\frac{dw}{d\lambda} = \frac{1.2}{m} + O\left(\frac{1}{m^3}\right), \quad \delta w = \frac{0.8}{m} + O\left(\frac{1}{m^3}\right). \tag{19}$$

Formulas (17)–(19) generalize and specify the ones represented in [10].

A dispersion equation in the case of a shell with liquid has the form (8). It can be rewritten in the following way:

$$\left.\frac{f(x)}{x}\right|_{x=\sqrt{w^2 c^2 - \lambda^2}} = \frac{1}{w^2(\lambda)} \frac{\Pi_{j=1}^{j=3}[w^2(\lambda) - w_j^2(\lambda)]}{\Pi_{j=4}^{j=5}[w^2(\lambda) - w_j^2(\lambda)]}, \tag{20}$$

where

$$f(x) \equiv -\frac{1}{M}\frac{J_m(x)}{\partial_x J_m(x)}; \quad M \equiv \sqrt{12}\,\alpha\rho^*,$$

$w_j(\lambda), j = 1, 2, 3$ are the roots of the Eq. (8) and $w_j(\lambda), j = 4, 5$ are the roots of the biquadratic equation $M_{3,3}(w^2, \lambda^2) = 0$ and can be represented in the asymptotic form

$$w_j^2(\lambda) = w_j^2 + B_j \lambda^2 + O(\lambda^4), \quad j = 4, 5,$$

where $w_4 = w_2$.

The bifurcation condition in this case has the following view:

$$\frac{f(x)}{x}\bigg|_{x=cw_2} - \frac{(w_2^2 - w_1^2)(w_3^2 - w_2^2)}{w_2^2(w_5^2 - w_2^2)} = 0. \tag{21}$$

Asymptotic expansion is searched for in the form (12). After substituting it into the dispersion equation (20), coefficients A and B are determined. Particularly,

$$A = \pm \sqrt{\frac{r}{s+t}}, \tag{22}$$

where

$$r = \frac{[B_2 - B_4][w_2^2 - w_1^2][w_3^2 - w_2^2]}{[w_5^2 - w_2^2]},$$

$$s = 1 + \frac{[w_5^2 - w_1^2][w_3^2 - w_5^2]}{[w_5^2 - w_2^2]^2},$$

$$t = -\frac{1}{2x}[xf(x)]\bigg|_{x=cw_2}.$$

Elementary analysis shows that $r, s, t > 0$ and the square root in (22) is determined. One can obtain the asymptotic view of group velocity using the representations

$$B_2 = v_- \left(1 + \frac{4v_+}{1 + m^2 v_+}\right) + O(\alpha^2),$$

$$B_4 - 1 - v_+^2 / (\alpha_1 - v_-) = v_+ + O(\alpha^2),$$

$$w_1 = \frac{m(m^2 - 1)}{\sqrt{1 + m^2}} \alpha + O(\alpha^2),$$

$$w_3 = \sqrt{1 + m^2} + O(\alpha^2),$$

$$w_5 = m\sqrt{\alpha_1} = m + O(\alpha^2)$$

and substituting them into (22). As a result,

$$\left.\frac{dw}{d\lambda}\right|_{\lambda=0} = \pm\frac{1}{m}\sqrt{\frac{2v_-}{g(cw_2)}}\alpha^{1/2} + O(\alpha^{3/2}), \tag{23}$$

where

$$g(x) \equiv \frac{1}{\sqrt{12}}\frac{1}{\rho x}\partial_x\left[\frac{xJ_m(x)}{\partial_x J_m(x)}\right] > 0; \quad x > 0. \tag{24}$$

As was mentioned above, all of the processes in the liquid and shell are supposed to be harmonic with frequency ω. It is convenient to average the energy streams over a period of oscillations $T = 2\pi/\omega$. The integral energy stream Υ in the liquid along axes z through the cross-section of the cylinder can be written in the form

$$\Upsilon = \frac{\omega}{2}\frac{1}{\rho_w\omega^2}\int_0^{2\pi}d\varphi\int_R^{+\infty}\text{Im}\left(\overline{P}\frac{\partial P}{\partial z}\right)rdr. \tag{25}$$

The integral stream of the energy along axes z through the cross-section of the cylinder shell looks like this:

$$\Pi = \frac{\omega}{2}\int_0^{2\pi}\text{Im}\left(\mathbf{u}^4, \mathbf{F}\,\mathbf{u}^4\right)_{C^4}Rd\varphi = \Pi_t + \Pi_z + \Pi_n + \Pi_p, \tag{26}$$

where $\mathbf{u}^4 = (u_t, u_z, u_n, u_p)^t$, $u_p = -\partial_z u_n$ is the vector of generalized displacement and $\mathbf{F}$ is the matrix differential operator 4×4 [8]. Here, letters t, z, n and p mark the tangential (rotating), longitudinal, normal and momentum components of energy flux Π and the components of the vector of generalized displacement $\mathbf{u}^4$.

In the particular case of axisymmetric rotating movements of the cylindrical shell, the integral energy flux in it looks like this:

$$\Pi^0 = 2\pi\rho c_s^2\frac{\omega}{2}|A|^2\beta \equiv \Pi_t^0; \quad \beta = w\sqrt{(1+4\alpha^2)(1-v)/2}. \tag{27}$$

Formulas (25), (27), (28) can be used to obtain the normalized energy stream S in the shell and its components $S_{t,z,n,p}$, and to obtain the normalized energy stream W in the liquid

$$S = \Pi/\Pi^0, \quad S_{t,z,n,p} = \Pi_{t,z,n,p}/\Pi^0, \quad W = \Upsilon/\Pi^0. \tag{28}$$

Numerical Calculations

Calculations were fulfilled for the second mode ($m = 2$) and for the following values of parameters of the system $v = 0.28$, $\rho = 7.8$, $h/R = 0.0192$.

Figure 1a illustrates the case of dispersion curve bifurcation. The starting points of the second and third dispersion curves coincide. As a result, the right branch (marked by the letter "c") and the left branch (consisting of the section marked by the letter "b" and the section with negative group velocity, marked by the letter "a") were formed. The second dispersion curve of the "dry" shell is marked by the letter "d". One can see that the group velocity at the point of bifurcation has linear asymptotic on λ and equal values of module of group velocity for the left and right branches.

Figure 1b illustrates Eq. (21). The first summand consists of c and ρc parameters characterizing the shell-liquid interaction. The second summand of this equation depends on the shell parameters alone. Roots c_j, $j = 1, 2, 3, \ldots$ of this equation can be easily analyzed from the graph of the left side of expression (21) as a function of c. This graph has an infinite number of zeroes (roots of Eq. (21)) between vertical asymptotes at the points where

$$\partial_x J_m(x)|_{x=cw_2} = 0. \tag{29}$$

These roots c_j (and correspondingly, $x_j = c_j w_2$) are going to the roots of the dispersion equation of the liquid cylindrical waveguide with rigid boundaries (29) when the number "j" increases. Hence, t in (22) increases too, and $dw/d\lambda$ vanishes. This fact is in good accordance with Fig. 2a. In this figure, the dependence of the dispersion curve section with the negative group velocity for four different relative velocities $c = 3.60, 6.13, 8.69, 11.27$ (curves $1, 2, 3, 4$) is shown. The dash—dotted curve B is the second curve of the "dry" shell. If the shell is relatively softer (c is smaller), then the section with negative group velocity is greater.

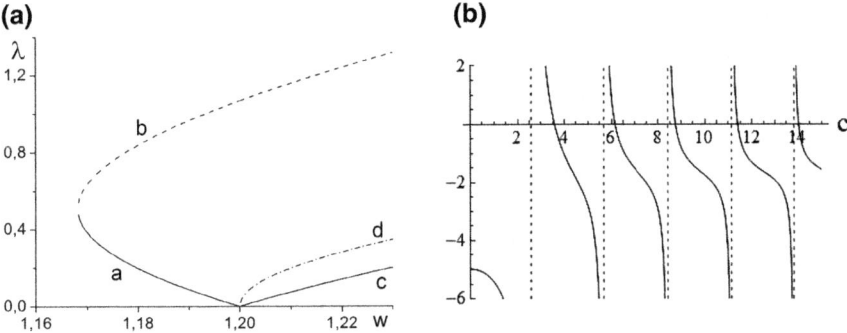

Fig. 1 **a** The dispersion curves bifurcation; **b** the bifurcation condition

The wavenumber intervals where the negative group velocity is observed $\delta\lambda < 0.5$ are less then in the case of a dry shell $\delta\lambda < 4/3$ (18), due to the adjoint mass of liquid in the first case.

The normalized energy fluxes in the liquid (W) and the shell (S) and the normalized rotating, longitudinal, bending and momentum components ($S_{t,z,n,m}$) of the energy flux in the shell (28) illustrate the wave processes in the shell-liquid system of the next figures. The energy fluxes of the waves from the left dispersion curve are represented for the frequencies less the bifurcation one alone. The letters "a", "b", "c" in the designations of these curves conform to the corresponding sections of the dispersion curves in Fig. 1a.

In Fig. 2b, curves 1, 2, 3 correspond to the energy fluxes W, S, $S + W$. The energy flux in the shell is dominant. It is not equal to zero compared to the energy flux in the liquid, and has the opposite sign and the equal module when $\lambda \to 0$. The energy flux in the liquid remains positive and is equal to the module for the waves from both branches in the neighborhood of the bifurcation point.

Distribution of the energy flux to different components is shown in Fig. 3, where curves 1 and 2 correspond to the energy flux components S_t and S_z. The main role

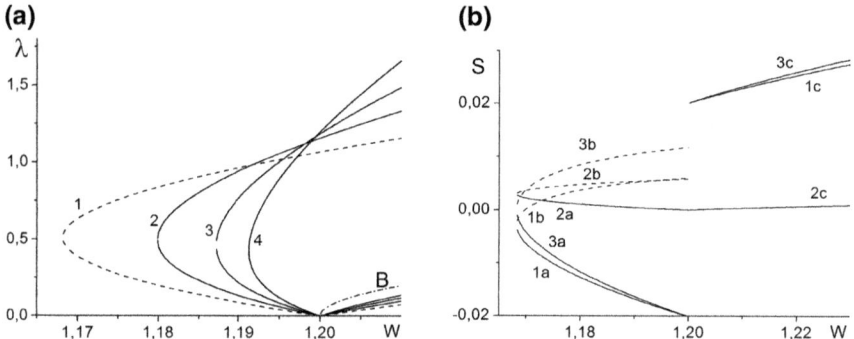

Fig. 2 **a** The dispersion curves for different c; **b** the energy fluxes in the shell (1), in the liquid (2) and their sum (3) for the bifurcating dispersion curves

Fig. 3 The energy flux S in the shell and its components S_t (1), S_z (2) for the bifurcating curves

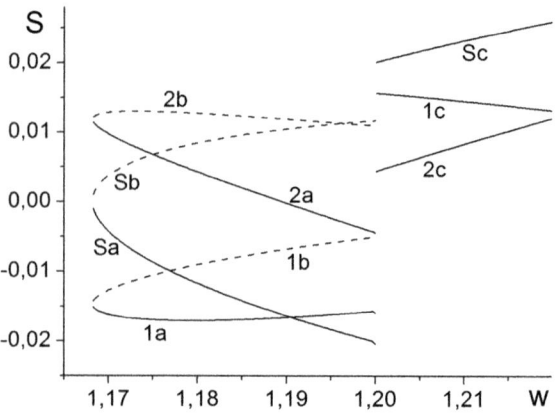

in the energy transmission belongs to the rotating and longitudinal components of an energy flux, with the first one dominating. Calculations show that the limiting values of the bending and momentum components S_n and S_m are equal to zero at the bifurcation point and are much less than the rotating and longitudinal components in its neighborhood.

Conclusions

The analysis carried out shows that the Kirchhoff–Love shell model of the "dry" shell considered as an abstract (mathematical) model allows us to qualitatively simulate the essential effects in the more difficult "shell-liquid" system. In particular, it shows that the specific character of the waves with negative group velocity at the point of bifurcation is their quick switching on the long waves in the energy transmission process (the limiting values of group velocity and energy flux are not equal to zero at this point).

In the case of a shell with liquid, the negative energy flux of the shell determines the negativeness of the whole energy flux in the neighborhood of the bifurcation point. The main role in energy transmission belongs to the rotating and longitudinal components of the shell energy flux with the first one dominating.

The relative "softness" of the liquid (parameter c) strongly affects the frequency range where the negative group velocity is observed.

The appearance of negative group velocity can be important for the nonhomogeneous problem. If the source vibrations' frequency is located near the starting point of the dispersion curve with a negative group velocity section, then the excitation coefficient of the corresponding wave can be significant. It can affect the strength characteristics of the shell.

References

1. Zinovieva TV (2017) Calculation of shells of revolution with arbitrary meridian oscillations. In: Evgrafov A (ed) Advances in mechanical engineering. Lecture notes in mechanical engineering. Springer International Publishing, Switzerland, pp 165–176. https://doi.org/10.1007/978-3-319-53363-6-17
2. Zinovieva TV (2016) Calculation of forced oscillations of shells of revolution with arbitrary meridian. In: Modern engineering: science and education. Proceedings of fifth international scientific and practical conference. State Polytechnic University, Saint Petersburg, pp 442–452. http://www.mmf.spbstu.ru/mese/2016/442-452.pdf
3. Yeliseyev VV, Zinovieva TV (2012) Nonlinear-elastic strain of underwater pipeline in laying process. Vycisl Meh Splos Sred—Comput Continuum Mech 5(1):70–78
4. Zinovieva TV (2007) Wave dispersion in cylindrical shell, vol 504. Acta of SPb SPU Engineering, St. Petersburg State University Publishers, pp 112–119
5. Yeliseyev VV, Zinovieva TV (2014) Two-dimensional (shell-type) and three-dimensional models for elastic thin-walled cylinder. PNRPU Mech Bull 3:50–70

6. Ter-Akopyants GL (2015) Axisymmetrical wave processes in cylindrical shells filled with fluid, Estestvennye-i-tehnicheskie-nauki (Nat Eng Sci) 85(7):10–14
7. Filippenko GV (2017) Energy-flux analysis of the bending waves in an infinite cylindrical shell filled with acoustical fluid. In: Evgrafov A (ed) Advances in mechanical engineering. Lecture notes in mechanical engineering. Springer International Publishing, Switzerland, pp 57–64. ISSN 2195-4356. https://doi.org/10.1007/978-3-319-53363-6
8. Filippenko GV (2016) The vibrations of reservoirs and cylindrical supports of hydro technical constructions partially submerged into the liquid. In: Evgrafov A (ed) Advances in mechanical engineering. Lecture notes in mechanical engineering. Springer International Publishing, Switzerland, pp 115–126. ISSN 2195-4356. https://doi.org/10.1007/978-3-319-29579-4
9. Filippenko GV (2016) The energy flux analysis of the "shell" type waves in the infinite cylindrical shell filled with acoustical fluid. In: Proceedings of the international conference "Days on Diffraction 2016", Saint Petersburg, Russia, pp 54–58
10. Filippenko GV (2017) Energy analysis of waves with negative group velocity in cylindrical shell. Vycisl Meh Splos Sred—Comput Continuum Mech 10(2):187–196
11. Filippenko GV (2014) Energy aspects of wave propagation in an infinite cylindrical shell fully submerged in liquid. Vycisl Meh Splos Sred—Comput Continuum Mech 7(3):295–305
12. Filippenko GV (2013) Energy aspects of asymmetrical waves propagation in the infinite cylindrical shell fully submerged into the liquid. Vycisl Meh Splos Sred—Comput Continuum Mech 6(2):187–197
13. Filippenko GV (2011) The energy analysis of shell-fluid interaction. In: Proceedings of the International Conferrence on "Days on Diffraction 2011", Saint-Petersburg, Russia, May 30–June 3, pp 63–66
14. Cremer L, Heckl M, Petersson, Bjorn AT (2005) Structure-borne sound. Structural vibrations and sound radiation at audio frequencies, 3rd edn, XII, 607 p, 215 illus
15. Fuller CR, Fahy FJ (1982) Characteristics of wave propagation and energy distributions in cylindrical elastic shells filled with fluid. J Sound Vib 81(4):501–518
16. Pavic G (1990) Vibrational energy flow in elastic circular cylindrical shells. J Sound Vib 142(2):293–310
17. Pavic G (1992) Vibroacoustical energy flow through straight pipes. J Sound Vib 154(3):411–429
18. Feng L (1994) Acoustic properties of fluid-filled elastic pipes. J Sound Vib 176(3):399–413
19. Sorokin SV (2011) The Green's matrix and the boundary integral equations for analysis of time-harmonic dynamics of elastic helical springs. J Acoust Soc Am 129(3):1315–1323
20. Sorokin SV, Nielsen JB, Olhoff N (2004) Green's matrix and the boundary integral equations method for analysis of vibrations and energy flows in cylindrical shells with and without internal fluid loading. J Sound Vib 271(3–5):815–847
21. Kouzov DP, Mirolubova NA (2012) Local energy fluxes of forced vibrations of a thin elastic band. Vycisl Meh Splos Sred—Comput Continuum Mech 5(4):397–404
22. Veshev VA, Kouzov DP, Mirolyubova NA (1999) Energy flows and dispersion of the normal bending waves in the X-shaped beam. Acoust Phys 45(3):331–337

On Equally Stressed Hinged Devices

Mikhail D. Kovalev

Abstract The paper addresses the search for flat hinged-lever structures with the same kinematic scheme, allowing for the same complete self-stress. If all of the fastened hinges lie on a straight line, then they are certainly mirror-symmetric with respect to a straight line construction. In the case of two infinite sequences of structural schemes with non-collinear fastened hinges, the absence of trusses with the desired properties is proved. The restriction on the rank of the stress matrix establishes a theorem on the structure of the hinged mechanisms that allow for the same complete self-stress in each position.

Keywords Graph design in a plane · Fastened hinged scheme
Kinematic scheme · Self-stress · Uniqueness of assembly

Introduction

We consider the geometric and static properties of ideal flat hinged-lever constructions. Since the article is of a mathematical nature, it is necessary to define a number of concepts accurately for a better understanding. We will now do this, and at the end of the introduction, we will briefly describe the results set out in the remaining sections.

We call a *hinged structural scheme* (HSS, for short) a connected graph $G(V, W, E)$ without loops and multiple edges, whose vertex set consists of two nonempty subsets $V = \{v_1, ..., v_m\}$ (in our figures, they are denoted as circles, corresponding to *free hinges*) and $W = \{v_{m+1}, ..., v_{m+n}\}$ (these are crosses, corresponding to *fastened hinges*). Graph $G(V, W, E)$ satisfies the following conditions: its subgraph $G_1(V, E_1)$ generated by the set V is connected, and the vertices of the set W are not adjacent to each other [1, 2]. Thus, the set E of edges of the graph $G(V, W, E)$ is split into two subsets: E_1, consisting of edges $v_i v_j$, for which $i, j \leq m$; and E_2,

M. D. Kovalev (✉)
Moscow State University, Moscow, Russia
e-mail: kovalev.math@mtu-net.ru

© Springer International Publishing AG 2018
A. N. Evgrafov (ed.), *Advances in Mechanical Engineering*, Lecture Notes
in Mechanical Engineering, https://doi.org/10.1007/978-3-319-72929-9_12

consisting of edges $v_i v_j$, with $i \leq m$, $j > m$. A *fastened hinged scheme* (FHS) is called an HSS with each of its fastened hinges $v_i \in W$ mapped to some point p_i of the plane. Moreover, different points are mapped to different hinges. If we now map each free hinge to a point of the plane, we get *a framework* $p = (p_1, p_2, \ldots, p_m)$ corresponding to the given FHS. In our model, the edges of the graph correspond to the levers connecting the hinges of the framework. We do not exclude the coincidence of various hinges as points of the plain and the intersection of different levers. A framework can be thought of as either a truss or a definite position of the hinged mechanism. Mathematically, a framework is a point of a $2m$-dimensional Euclidean space of coordinates of free hinges $p \in R^{2m}$.

Another parameter space, the r-dimensional Euclidean space R^r, whose coordinates are the squares of the lengths of the levers, also corresponds to a hinged structural scheme with r levers. Each point of this space $d = \{d_{ij}\}$, $v_i v_j \in E$ is called the *kinematic hinged scheme* (KHS). A given FHS determines the mapping between the introduced spaces of parameters $F : R^{2m} \rightarrow R^r$, by the formulas $d_{ij} = (p_i - p_j)^2$, $v_i v_j \in E$. This mapping is called *rigidity mapping* [3–5]. It plays a key role in the geometric analysis of hinged constructions.

The matrix dF of the differential of the rigidity mapping plays an important role in the study of hinged constructions. In the case of flat structures, this matrix is of size $r \times 2m$, with the rows corresponding to the levers of the HSS. The row corresponding to the lever $p_i p_j$, connecting the free hinges, looks like

$$2(0, \ldots, 0, x_i - x_j, y_i - y_j, 0, \ldots, 0, x_j - x_i, y_j - y_i, 0, \ldots, 0),$$

where $p_i = (x_i, y_i)$, and all of the remaining elements of the row except the four differences are equal to zero. The row of the matrix dF, corresponding to the lever $p_i p_j$, where $v_i v_k \in E_2$, looks like

$$2(0, \ldots, 0, x_i - x_k, y_i - y_k, 0, \ldots, 0, 0, \ldots, 0).$$

If Rank $dF(p) = r$, then framework p is called *regular*; otherwise, it is *irregular*. In the case of Rank $dF(p) = r = 2m$, framework p is a statically determinate truss. We call it *isostatic*. An FHS is called *regular* (*isostatic*) if it corresponds to at least one regular (isostatic) framework. In mechanics, this corresponds to the absence of passive links in the construction. A necessary and sufficient condition for the regularity and isostaticity of an HSS in the plane is given in [1].

The linear dependence of the rows of the matrix $dF(p)$ with coefficients ω_{ij} has a static meaning. Namely, it is equivalent to the fulfillment, for ω_{ij}, of the system of vector linear equations:

$$\sum_j \omega_{ij}(p_i - p_j) = 0, \quad 1 \leq i \leq m, \tag{1}$$

where the summation is performed over all of the hinges adjacent to p_i. These equations are the conditions for the equilibrium of forces applied to i-th free hinge from the adjacent hinges. The values $\omega_{ij} = \omega_{ji}$ of the *internal stresses* of the levers indicate the measure of the tension of the levers: if $\omega_{ij} < 0$, then the lever $p_i p_j$ is stretched, if $\omega_{ij} > 0$, then it is compressed.

In the engineering literature, the term "tension of the lever" has a different meaning. Namely, the tension is the force per unit area of the cross-section of the lever.

If system (1) has only a trivial solution, then it is said that the framework does not allow *self-stresses*. Otherwise, the set $\omega = \{\omega_{ij}\}$ of self-stresses of the framework, together with the trivial solution to the system (1), represents the *linear space* $W(p)$ *of the self-stresses* of the framework.

Consider a framework p having a self-stress $\omega = \{\omega_{ij}\}$. We rewrite system (1) as a system of equations with respect to the radius vectors of the free hinges, assuming that the self-stress ω is specified:

$$\left(\sum_{(i,j)\in E} \omega_{ij} \right) p_i - \sum_{(i,j)\in E_1} \omega_{ij} p_j = \sum_{(i,j)\in E_2} \omega_{ij} p_j, \quad 1 \leq i \leq m. \qquad (2)$$

We emphasize that on the right-hand side of the equations of this system, there are certain radius vectors of fastened hinges. The set of solutions of this system is either a point or a *linear manifold* L_ω of the dimension multiple of two [6]. The manifold L_ω consists of frameworks with a given FHS, allowing for self-stress $\omega = \{\omega_{ij}\}$. In the case of the uniqueness of the solution to system (2), we will speak of the *restorability of the framework p from its self-stress* ω [7, 8]. An important role is played by the matrix $\Omega(\omega)$ of system (2), called the *stress matrix* [6, 9]. This is a symmetric $m \times m$-matrix with elements:

$$\Omega_{ij} = 0, \quad (i,j) \notin E_1, \quad \Omega_{ij} = -\omega_{ij}, \quad (i,j) \in E_1, \quad \Omega_{ii} = \sum_k \omega_{ik},$$

where the last sum is taken over all levers adjacent to the i-th hinge.

We will investigate the following questions: under what conditions are there no isometric frameworks among frameworks with a given FHS, admitting a given self-stress ω? Isometric frameworks are strictly frameworks with the same KHS.

In the next section, we give an example of a manifold L_ω and an example of a framework that is unrestorable from a self-stress, as well as some preliminaries. We call an FHS with all fastened hinges lying on a straight line *straightened* [10], otherwise, we call it *non-straightened*. Seeking the conditions under which there are no isometric frameworks, it is reasonable to limit ourselves to considering self-stresses nonzero on each lever. We will call these stresses *complete*, and a framework that allows it, *completely stressed*.

Section "Stressed Isometric Frameworks with Schemes D_m and M_m" contains a theorem, establishing the absence of isometric, completely stressed frameworks for two infinite sequences of isostatic non-straightened FHS.

In Section "Isometric, Completely Stressed Trusses with Non-straightened Fastening", we give an example of isostatic non-straightened FHS, for which different isometric, equally and completely stressed frameworks exist. A theorem on frameworks that either have or do not have an isometric framework in L_ω, is formulated with the assumption that coRank $\Omega(\omega) = 1$.

Section "On Mechanisms Allowing Complete Self-stress" is devoted to hinged mechanisms that allow the same complete self-stress in each position. A theorem on the structure of such mechanisms is given in the case of coRank $\Omega(\omega) = 1$.

Preliminaries and Examples

In the planar case, the vector system (2) is split into two coordinate systems with identical matrices, but different right-hand sides. As follows from the Kronecker-Capelli theorem, a necessary and sufficient condition for the compatibility of these systems, and hence of system (2), is the equality of the ranks of the matrix $\Omega(\omega)$ and the twice expanded matrix $\Omega^*(\omega)$, obtained by the addition of two columns of the free terms of coordinate systems to $\Omega(\omega)$. The following simple assertions are true [4, 8].

Let a framework $A(p)$ be obtained by an affine transformation from a framework p.

Proposition 1 *The frameworks p and $A(p)$ have the same space of self-stresses.*

Proposition 2 *If a framework is restorable from its self-stress, and its fastened hinges lie on a straight line, then its free hinges lie on the same line.*

Searching for cases in which the manifold L_ω does not contain isometric frameworks, it suffices to confine oneself to the non-straightened FHS. Moreover, it is reasonable to consider only complete self-stresses ω. Indeed, for example, for the FHS D_2 (Fig. 1), there are two different isometric, incompletely stressed frameworks p and p'. Here, the levers emanating from the hinge p_2, unlike the collinear

Fig. 1 Two isometric equally and incompletely stressed frameworks with non collinear fastened hinges

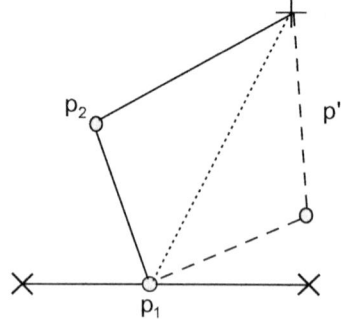

Fig. 2 An unusual set of frameworks allowing a given self-stress

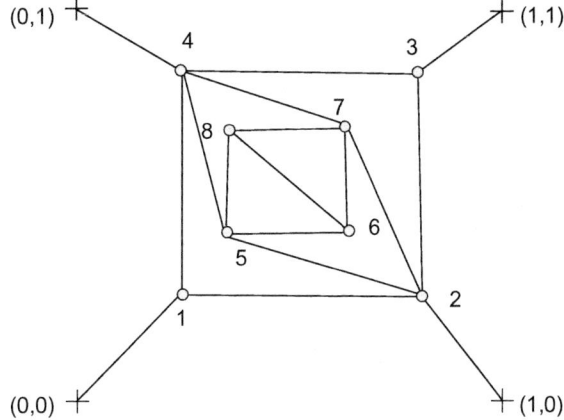

levers emanating from the hinge p_1, are not stressed. The framework p' is obtained from p by reflecting the part of the latter from the dotted line.

Example The irregular FHS of Fig. 2 with 8 free and four fixed hinges allows the following self-stress:

$$\omega_1 = 1, \quad \omega_{12} = 1, \quad \omega_{14} = -2, \quad \omega_2 = -7/6, \quad {}_{23} = -3, \quad \omega_{25} = -1/6, \quad \omega_{27}$$
$$= 1, \quad \omega_3 = 1, \quad \omega_{34} = -1,$$

$$\omega_4 = -7/3, \quad \omega_{47} = -2, \quad {}_{45} = 1/3, \quad \omega_{56} = 1, \quad {}_{58} = -1, \quad \omega_{67} = 1, \quad \omega_{68}$$
$$= -1, \quad \omega_{78} = -2,$$

(stress ω_i corresponds to the lever emanating from the i-th free hinge to the fastened one). For this stress, we have coRank $\Omega(\omega) = 2$. Here, the manifold L_ω depends on two parameters and is defined by the equations $p_1 = \frac{7}{3}p_4 + \frac{1}{6}\begin{pmatrix} 1 \\ -6 \end{pmatrix}$, $p_2 = 2p_4$, $p_3 = \frac{7}{3}p_4 - \frac{1}{3}\begin{pmatrix} 1 \\ 1 \end{pmatrix}$, and $p_5 = 2p_8$, $p_6 = \frac{4}{3}p_8$, $p_7 = \frac{1}{3}p_8$. The hinges fall into two groups. In the manifold L_ω, the position of the hinges of the inner square (with numbers 5, 6, 7, 8) depend only on the position of one of them (p_8), and do not depend on the positions of the hinges of the outer square.

Conversely, the positions of the hinges of the outer square depend only on the position of the hinge p_4.

The matrix $\Omega^*(\omega)$ in this case has the form

$$\begin{bmatrix} 0 & -1 & 0 & 2 & 0 & 0 & 0 & 0 & 0 & 0 \\ -1 & -\frac{7}{3} & 3 & 0 & \frac{1}{6} & 0 & -1 & 0 & -\frac{7}{6} & 0 \\ 0 & 3 & -3 & 1 & 0 & 0 & 0 & 0 & 1 & 1 \\ 2 & 0 & 1 & -7 & -\frac{1}{3} & 0 & 2 & 0 & 0 & -\frac{7}{3} \\ 0 & \frac{1}{6} & 0 & -\frac{1}{3} & \frac{1}{6} & -1 & 0 & 1 & 0 & 0 \\ 0 & 0 & 0 & 0 & -1 & 1 & -1 & 1 & 0 & 0 \\ 0 & -1 & 0 & 2 & 0 & -1 & -2 & 2 & 0 & 0 \\ 0 & 0 & 0 & 0 & 1 & 1 & 2 & -4 & 0 & 0 \end{bmatrix}.$$

Stressed Isometric Frameworks with Schemes D_m and M_m

We consider two infinite series of isostatic FHS: D_m and M_m (Fig. 3). The scheme D_m consists of the chain $p_1, p_2, \ldots, p_m$, composed of free hinges and the fastened hinges $q_0, q_1, \ldots, q_m$ attached to the chain. The fastened hinges q_0, q_1 are attached to the free hinge p_1, and the hinge q_i, $i > 1$ is attached to the hinge p_i.

The scheme M_m consists of the closed polygon $p_1, p_2, \ldots, p_m$, composed of free hinges, and the fastened hinges $q_1, \ldots, q_m$ attached to the polygon. The fastened

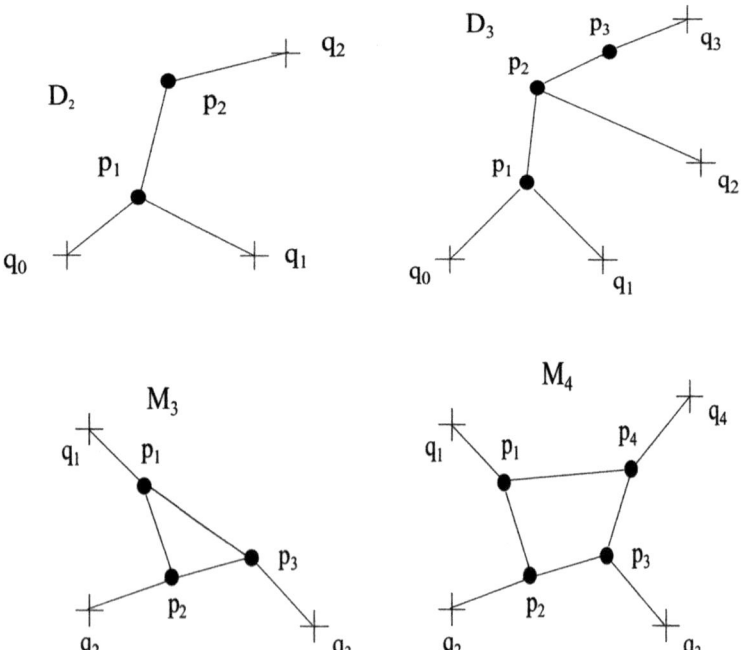

Fig. 3 Hinged schemes D_2, D_3, M_3, M_4

hinge q_i is connected to the hinge p_i. A framework is called *irreducible* if all of its levers have nonzero lengths.

Theorem 1 *For non-straightened isostatic schemes D_m and M_m, there is no KHS d, for which the full inverse image $F^{-1}(d)$ contains irreducible frameworks $p \neq p'$ admitting the same complete self-stress ω.*

Isometric, Completely Stressed Trusses with Non-straightened Fastening

However, there are frameworks with a non-straightened FHS, allowing complete self-stress, and having an isometric framework with the same self-stress. Figure 4 shows such an example with isostatic FHS. To the vertices of an equilateral rigid triangle p_4, p_5, p_6, parallel balanced forces are applied from the hinges p_1, p_2, p_3, each of which is adjacent to two fastened hinges, lying on a straight line parallel to the horizontal height of the triangle. If we make a parallel transfer of all free hinges as a rigid whole, then the balance of forces in each free hinge of the framework will remain. When moving free hinges up and down along the same distance, we obtain isometric frameworks.

If there are isometric frameworks in the manifold L_ω, the following theorem holds.

Theorem 2 *Suppose that $\Omega(\omega) = 1$, and there are at least two isometric frameworks in the manifold L_ω. Then, the set of frameworks in L_ω, having an isometric framework in L_ω, is dense everywhere in L_ω, and the set of frameworks in L_ω, having no isometric in L_ω, is, at most, one-dimensional.*

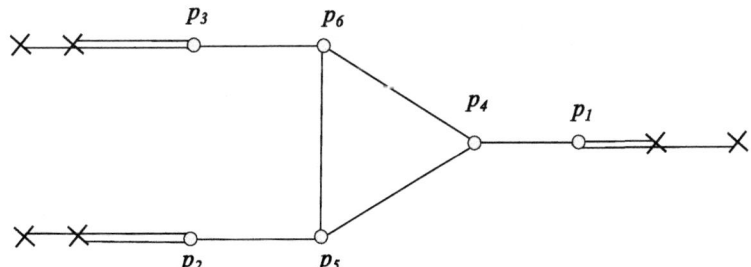

Fig. 4 A non-straightened FHS, for which exist equally and completely stressed isometric frameworks

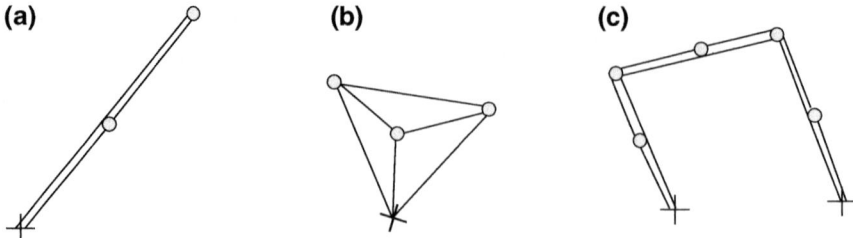

Fig. 5 Stressed mechanisms

On Mechanisms Allowing Complete Self-stress

There are not only trusses, but also hinged mechanisms that allow the same and complete self-stress in each position. Figure 5 shows three such examples. For the mechanism of Fig. 5a, we have coRank $\Omega(\omega) = 1$, and for the mechanisms of Fig. 5b, c, respectively, coRank $\Omega(\omega) = 2$.

Theorem 3 *A planar, irreducible hinged mechanism, assuming in each position a complete self-stress ω, for which* coRank $\Omega(\omega) = 1$, *and for which, for at least one position p of it,* dim $W(p) = 1$, *has a single fastened hinge, and all of its hinges lie on a straight line.*

The case described in the theorem can be seen in Fig. 5a. There remains an open question: are there such hinged mechanisms that do not contain parts larger than a lever moving like a rigid whole?

References

1. Kovalev MD (1994) Geometric theory of hinged devices. Izvestiay RAN Seriya Matematicheskaya 58(1):45–70
2. Kovalev MD (2001) Questions of the geometry of hinged devices and schemes. Vestnik MGTU. Seriya Mashinostroenie 4:33–51 (in Russian)
3. Connelly R (1993) Rigidity chapter 1.7. In: Gruber PM, Wills JM (eds) Handbook of convex geometry, vol A. Elsevier
4. Graver J, Servatius B, Servatius H (1993) Combinatorial rigidity. American Mathematical Society, Providence
5. Asimov L, Roth B (1979) The rigidity of graphs. II. J Math Anal Appl 68(1):171–190
6. Kovalev MD (1997) On the reconstructibility of frameworks from self-stresses. Izvestiya Mathematics 61(4):717–741
7. Kovalev MD (2013) A restoring stress doesn't always exist. In: Mechanism and machine science. In: Fernando V, Marco C (eds) New trends in mechanism and machine science, vol 7, pp 53–61. Springer, Dordrecht

8. Kovalev MD (2015) Frameworks not restorable from self-stresses. In: Evgrafov A (ed) Advances in mechanical engineering. Lecture notes in mechanical engineering. Springer International Publishing, Switzerland, pp 1–7
9. Kovalev MD (2016) The determinant of the stress matrix and restorability of hinged frameworks from self-stresses. Izvestiya Mathematics 80(3):500–522
10. Kovalev MD (2004) Straightened hinged frameworks. Sbornik Math 195(6):833–858

Determination of a Pre-destructive State During Hydraulic Testing of Steel Pipes with Defects by the Acoustic-Emission Method

Evgeny J. Nefedyev, Victor P. Gomera and Anatoly D. Smirnov

Abstract The results of hydraulic tests on the destruction of pipes made of 17G1S steel with a size of 630×10 mm and 219×6 mm made by welding high-frequency currents with artificial defects are presented. For the 630×10 mm pipe containing 3 half-elliptical crack-like defects of 100 mm in length, strength analysis was performed according to the finite element method in the ANSYS software package. The dynamics of accumulation of the damages and the development of the defects was controlled through the method of acoustic emission (AE). The criterion for estimating the degree of danger of the defects according to AE testing data is proposed. Good agreement was obtained when comparing the results of strength calculation and AE testing with the results of destructive tests.

Keywords Steel pipes · Crack-like defects · Hydraulic tests on destruction
Acoustic emission method · Finite element method · Criterion of pre-destructive state

Introduction

The AE method is actively used for investigation of structural material properties [1–11]. The purpose of the considered experiments was to develop a method of pre-destructive state determination and to evaluate a risk level of flaws in pipes made of 17G1S steel as a result of welding with high-frequency currents (HFC). The process of solution of the set task included the following stages:

E. J. Nefedyev (✉)
Central Boiler and Turbine Institute (CKTI), Saint Petersburg, Russia
e-mail: ne246@ya.ru

V. P. Gomera · A. D. Smirnov
Kirishinefteorgsintez, 187110 Kirishi, Russia
e-mail: Gomera_V_P@kinef.ru

A. D. Smirnov
e-mail: Smirnov_A_D@kinef.ru

© Springer International Publishing AG 2018
A. N. Evgrafov (ed.), *Advances in Mechanical Engineering*, Lecture Notes
in Mechanical Engineering, https://doi.org/10.1007/978-3-319-72929-9_13

1. Strength calculation of a pipe with several artificial defects according to the finite element method with the help of ANSYS program.
2. Detection of the most dangerous developing defects by way of the acoustic emission (AE) method.
3. Correlation of strength calculation and AE testing results with the non-destructive examination results.
4. Development of flaw risk level criterion through the AE method.

Investigation Objects

Destructive tests with internal pressure during hydraulic testing (HT) were carried out at the test facility of OJSC "NPO CKTI". The investigation targets were as follows:

1. Model D14, representing a pipe with a diameter of 630 mm, length of 3000 mm and wall thickness of 10 mm, manufactured of steel grade 17G1S. The pipe was closed, with elliptical heads at both ends, welded by manual arc welding (Fig. 1a). Three stress concentrators were arranged for the target, simulating a semi-elliptical crack with a length of 100 mm and a depth of 8 mm.
2. Model D15, representing a pipe with a diameter of 219 mm, a length of 2200 mm and a wall thickness of 6 mm, manufactured of steel grade 17G1S (Fig. 1b). Two stress concentrators were arranged for the target, simulating a semi-elliptical crack with a length of 60 mm and a depth of 4.8 mm (Fig. 2a).

The CDAE-16 system produced by "Promdiagnostika" LLC was used for AE testing of the D14. 4 acoustic emission transducers (AET) of the AE-225 type were installed during the tests. Use was also made of the AMSY5 AE system produced by Vallen Systeme (Germany) with resonant AE sensors V150-RIC and a wideband

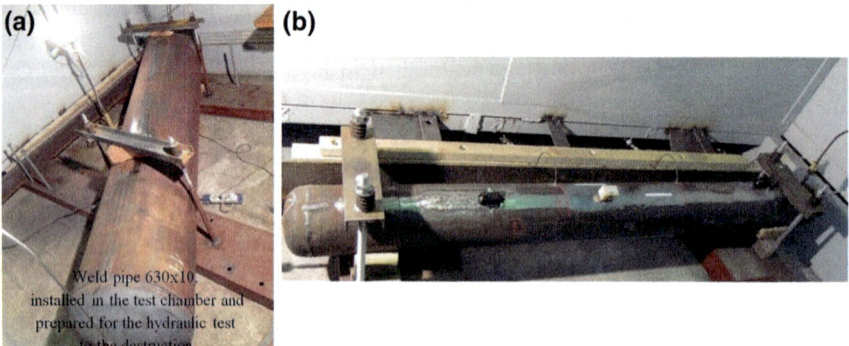

Fig. 1 **a** Hydraulic test scheme for model D14; **b** hydraulic test scheme and position of AE sensors for model D15

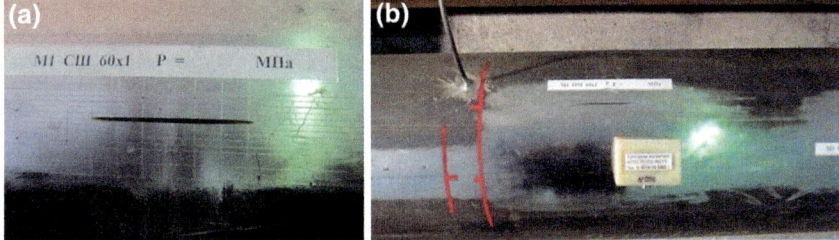

Fig. 2 **a** Stress concentrator in the form of a semi-elliptical crack 60 mm long and 4.8 mm deep on the surface of the tube 219 × 6 mm and **b** options for installing the AE sensors and their location relative to artificial defects: through the waveguide (on the left in the photo) and using the magnetic holder (in the center of the photo) on the D15

AE sensor WDI (PAC, USA). Figure 2 shows the AET arrangement on the test target for model D15.

Both installation options of AET, that in which contact is aided by waveguides and that in which the contact is done directly on the metal's surface through the use of magnetic holders, were applied (Fig. 2b).

Tests were carried out prior to creation of the through defects in the metal of the models. Figure 3 illustrates the crack opening and leak formation in the area of the artificial defect on the 210 × 6 mm pipe.

Strength Calculation Results for Model D14

It is possible to preliminarily evaluate a pressure p_c, at which 630 × 10 mm pipe destruction begins with a single axial defect with a depth of $a = 8$ mm and a length of $l = 100$ mm, using the simplified formula [12]:

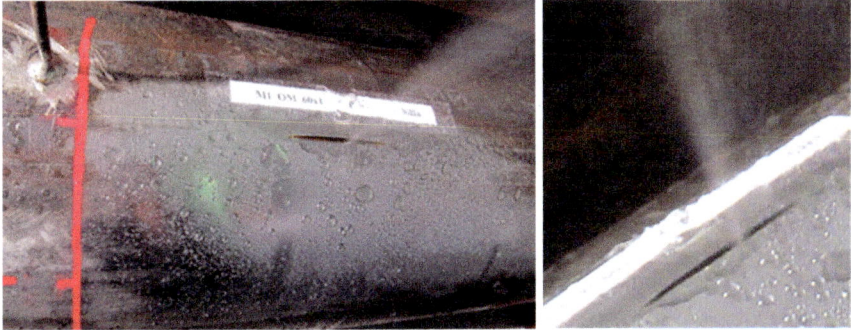

Fig. 3 Destruction (leakage of the medium) on the D15 model under pressure $P = 90$ kg/cm^2

$$p_c = \frac{2t}{D}\sigma_f(1 - a/t)/(1 - a/t/M_2), \quad M_2 = \sqrt{1 + (1.61/4RT)l^2}, \quad (1)$$

where $\sigma_f = \frac{\sigma_{0.2} + \sigma_B}{2}$.

The average yield strength is 460 MPa and the tensile strength is 580 MPa, according to data on pipe metal mechanical property measurement from the manufacturing plant. The critical pressure for a pipe with a single flaw, determined by the specified values according to formula (1), is 7 MPa. With account for possible scatter of the mechanical properties, the deviation of the actual values of destructive pressure from the design values can be up to 20%.

Let us give the design values (by average mechanical properties) for a flawless pipe as an example: limit pressure is 8.8 MPa, destructive pressure is about 11 MPa.

At the present time, the limit state evaluation for a structure with crack-like defects is based on the use of a two-criterion failure assessment diagram (FAD). This approach allows for determination of the critical load at the specified defect with account for material properties. This method is used in the foreign normative documents on designing pressure equipment [13, 14]. However, the scope of the existing normative FADs is limited to consideration of standard elements (pipe, bend) with single defects of postulated sizes (as a rule, $a/t < 0.5$). In a case of a cluster of defects, the conservative approach is offered, based on the replacement of several defects with one defect of equivalent size. That is why the method of critical load determination for a pipe with several defects at $a/t = 0.8$ (and plotting of the corresponding FAD) is based on calculation of the elastoplastic J integral and the limit load according to numeric methods (Fig. 4).

The specified calculation of FAD for a pipe with a system of three defects was carried out through the finite element method in the ANSYS software suit. The FAD was plotted as a dependence of $(J_{total}/J_{elastic})^{0.5}$ on a parameter $S_r = p/p_0$. The material stress-deformation diagram used in the calculation is given in Fig. 5. Represented in Fig. 6 are: the calculated FAD (curve 2) for the considered crack system, and the normative FAD (curve 1) [13] for the infinitely long axial defect. The calculated limit pressure value for the considered defect configuration was $p_0 = 5.8$ MPa.

Fig. 4 Section of a D14 pipe with a defect system

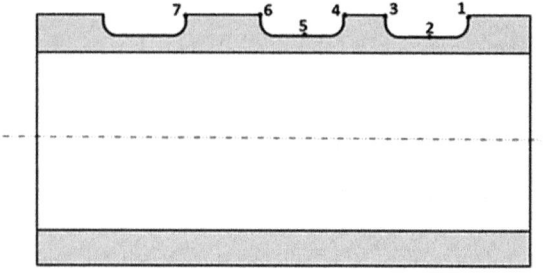

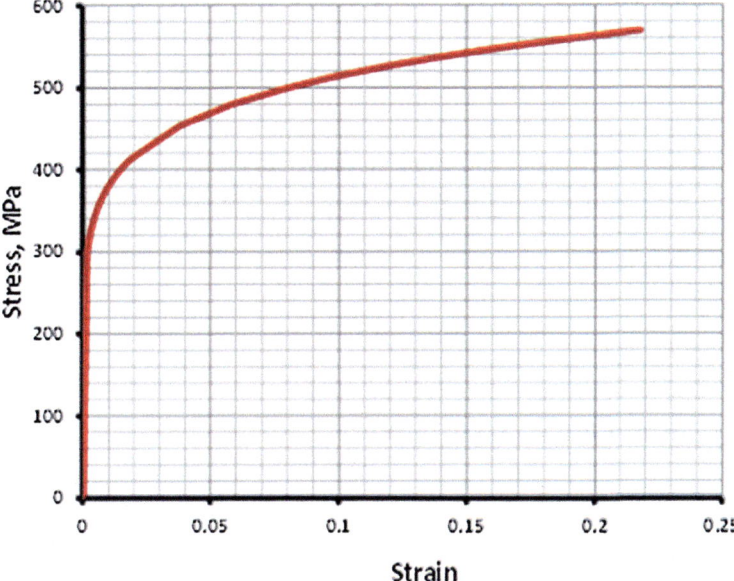

Fig. 5 Diagram of deformation of pipe material

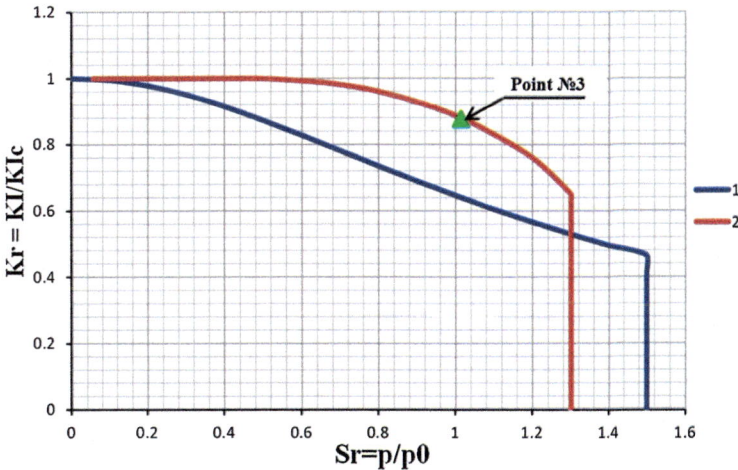

Fig. 6 FAD-diagrams

According to calculation results, the maximum value of stress-intensity factors (SIF) and, respectively, the parameter $K_r = K_I/K_{Ic}$ shall be achieved in the right-most defect at the front point of defect No. 3 (Fig. 4). It should be noted that the difference in SIF at the considered calculation points Nos. 1–6 was no more than

10%. Material viscosity was taken as per data [15] $K_{Ic} = 101$ MPa m$^{1/2}$. The destructive pressure predicted by calculation corresponds to the viscous area of destruction and is $P = 5.9$ MPa for the accepted material properties.

It follows from the calculation results that the pipe material through the whole of its cross-section has passed into a plastic state at the moment of total destruction. Since local bending stresses exceeding the pipe stress occur in the area of head welding to the pipe, a local flow area is developed there under a smaller degree of pressure than that in the pipe. It is quite possible that the occurrence of signals in the weld area is connected with the above-mentioned condition alone. But since these stresses are local and are not those of membranes (i.e., they are not uniformly distributed by cross-section), they cannot lead to total destruction.

Discussion of AE Testing Results

The AE activity graph (accumulation of signals) depending on the loading stage of model D14 is given in Fig. 7. Loading was carried out step-wise with a hold time of 5 min at each stage. Model destruction took place under HT pressure of $P = 120$ kg/cm^2. The accumulation of signals is of a power nature. 577 signals are registered in total. Figure 8 shows a diagram of the AE sensor positions on the reamer of the pipe (model D14) and the results of the AE sources' planar location.

The line-cutting of the reamer passes along the pipe's lower generating line. The weld has a coordinate of $Y = 1000$ mm. Coordinates of artificial concentrators: along the ordinate axis, $Y = 1000$ mm, boundaries along the abscissa axis, $X_{11} = 2170$ mm, $X_{12} = 2270$ mm (left), $X_{21} = 2370$ mm, $X_{22} = 2470$ mm

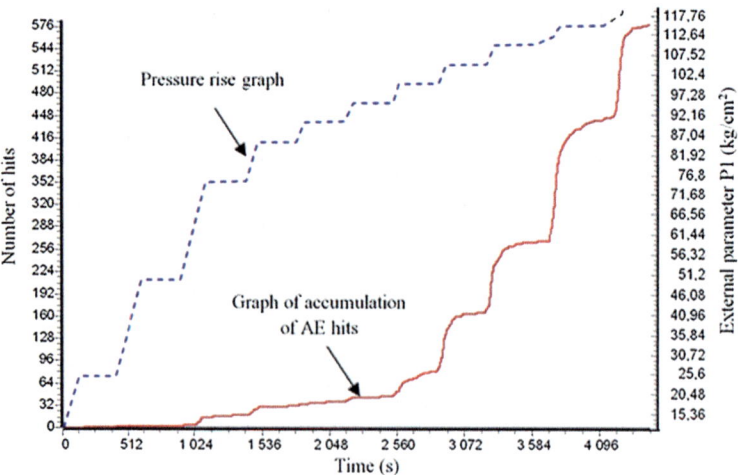

Fig. 7 Graphs of pressure rise and accumulation of the number of AE signals for model D14

(central), X_{31} = 2520 mm, X_{32} = 2620 mm (right). The position of the concentrators is shown in Fig. 8.

Let us consider the AE sources' distribution along the model surface or the so-called damage accumulation nature. Figure 8 shows the AE sources' distribution at three loading stages and over the whole test procedure.

Three AE sources in total are registered at the stage from P = 0 to P = 100 kg/cm². Two of them are within the area of right concentrator location.

At the stage from P = 100 to P = 110 kg/cm², occurrence of 7 additional AE sources is registered. At that, two of them are also within the area of the third cut.

At the last (third) loading stage, 29 localized AE sources are registered. In general, they are located within the area of the third cut and in the gap between the third and second cuts. Clusters of AE sources occur in the area of the welding of the right head.

While analyzing the general test results represented in Fig. 8, formation of several clusters—areas of AE sources compact location is observed.

Occurrence of AE sources in the area of the welding of the right head corresponds to the strength calculation, which predicts stress concentration in this area.

Important information about destruction is contained in the dependence of the AE signals' amplitude on load (Fig. 9). The graph shows that the first AE signal is registered under pressure of about P = 20 kg/cm². The most active emission is registered beginning from a pressure of 50–60 kg/cm². Signal amplitudes are maximal. They do not exceed 62 dB and do not increase with the pressure growth. This is indicative of the plastic nature of model destruction.

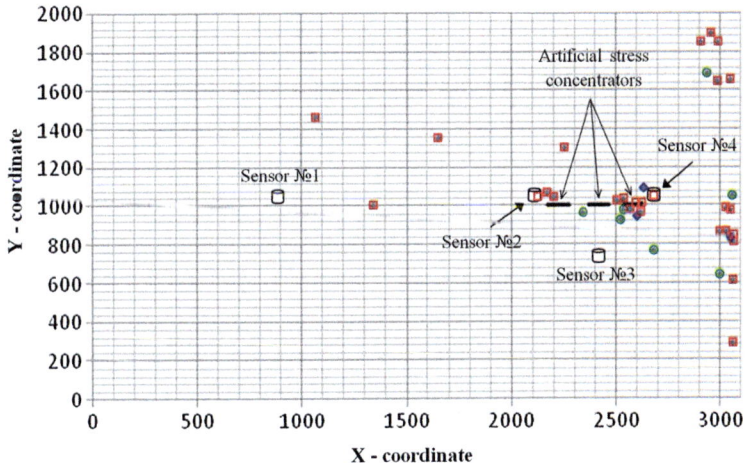

Fig. 8 The emplacement AE sensors on the model D14 and the results of the planar location. Designations of AE sources by registration time during hydraulic test: ◇—stage P = 0–100 kg/cm² (3 sources), ○—stage P = 100–110 kg/cm² (7 sources), □—stage P = 110–120 kg/cm² (29 sources)

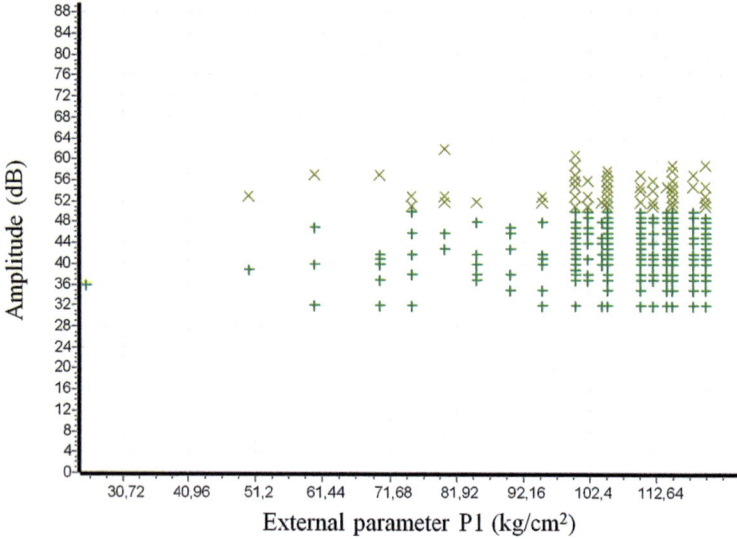

Fig. 9 Modification in the amplitude of AE signals as a function of the pressure of the hydraulic test

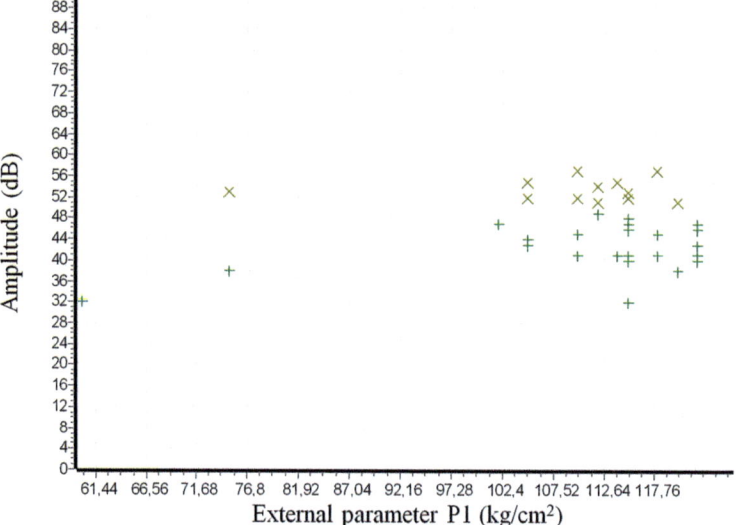

Fig. 10 Dependence of signals' amplitude of localized AE events on load

Figure 10 shows the dependence of AE localized signal amplitudes on pressure. Coordinates of 39 AE signals are determined in total. This is a normal relation between the total number of registered signals and the localized signals at the planar location of the events AE. The localized signals are registered beginning from the pressure of $P = 60$ kg/cm^2. The nature of dependence of the localized signals' accumulation on pressure is more abrupt than the nature of all AE signals' accumulation and is closer to the exponential law. Amplitude of these signals does not exceed 57 dB and does not increase with pressure growth.

In order to characterize the AE sources' distribution along the pipe surface, it is possible to calculate the concentration of the localized AE sources per unit area. The unit area is a portion of 200×200 mm. It is close to cut sizes. Besides, a number of AE sources different from "one" falls within such a portion. The whole pipe surface contains 150 such portions. Average concentration of AE sources is $C_{aver} = 39/150 = 0.26$ of AE sources per unit area. With such an approach, it is possible to distinguish several areas of AE activity:

(1) The area of the third cut (right), $C = 11$;
(2) The area of the first cut (left), $C = 3$.
    The remaining areas are in the right head weld. They can be identified by the Y-coordinate:
(3) $Y = 850$, $C = 6$; (4) $Y = 1000$, $C = 4$; (5) $Y = 1650$, $C = 3$; (6) $Y = 1850$, $C = 3$.

Comparing the value C for different activity areas and its dynamic pattern over the period of tests, it can be concluded that in the third concentrator area, formation of the most hazardous developing defect took place. The maximum concentration of AE sources is registered in this area. As a result, tests of model D14 have ended in destruction of the shell of this concentrator.

The validity of such an approach for determination of the area of the most probable destruction by concentration of AE sources can be demonstrated by the results of AE testing of model D15's destruction. The nature of mutual arrangement of defects and AET on D15 has allowed for performing both planar and linear location of AE sources using the same data set. The definite advantage of the linear location is the possibility of localizing a greater number of events. Figures 11 and 12 show the results of planar and linear location of AE events, respectively. Destruction took place under HT pressure equal to $P = 90$ kg/cm^2.

The graph of AE sources' distribution along the length of pipe generating the line (the solid line in Fig. 12) determines the position of the most hazardous flaw and the area of pre-destruction.

In order to compare the data obtained by different instruments, let us introduce a dimensionless factor of source concentration K_{AE} as a concentration of AE sources C_i normalized to the average AE source concentration for the whole pipe (C_{av}).

$$K_{AE} = C_i/C_{av} \qquad (2)$$

The physical sense of K_{AE} demonstrates how many times the AE source concentration in the given unit area exceeds the average concentration on the whole pipe surface.

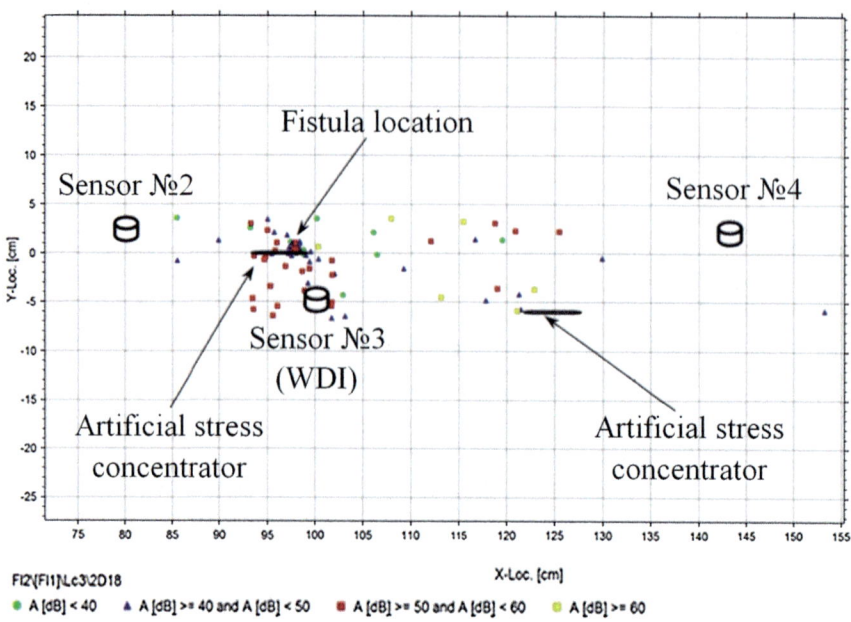

Fig. 11 The positions of AE sensors and artificial stress concentrators on the D15 model with the results of the planar location

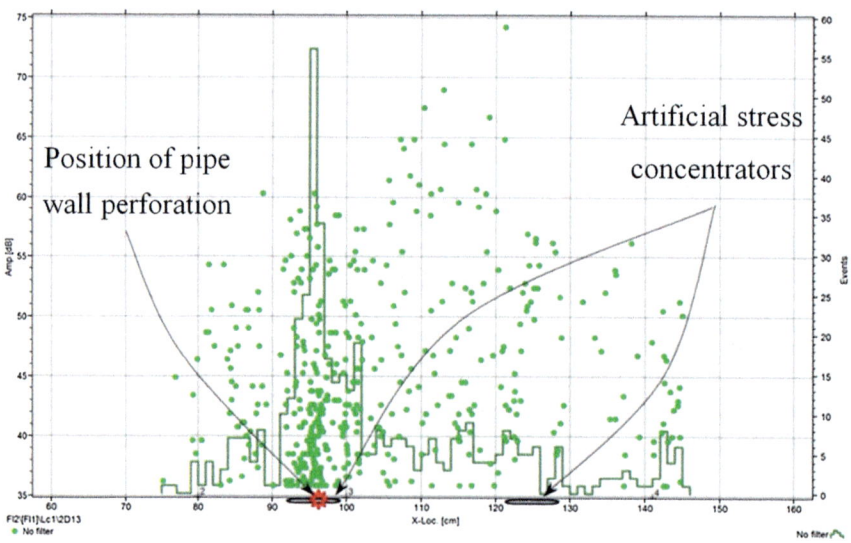

Fig. 12 The results of the linear location for the D15 model

Average concentration for a pipe with a diameter of 630 mm was $C_{av630} = 0.26$, for a pipe with a diameter 219 mm, $C_{av219} = 1.75$. Maximum concentration C_{max} was: for a pipe of 630, $C_{max630} = 11$, for a pipe of 219, $C_{max219} = 52$. These values, obtained with the destruction of different pipes, are fixed by different devices and differ by almost 5 times.

The normalized concentration of AE sources for a pipe with a diameter of 630 mm will be $K_{AE630} = 42$, and for a pipe with a diameter of 219 mm, $K_{AE219} = 30$. Thus, the concentration planar approach gives close values at the destruction of different objects and at application of different AE systems. This approach means that very high unevenness in the distribution of AE sources at the object should be recorded at the pre-destruction stage. The largest concentration of AE sources, tens of times higher than the mean values, should be observed in the region where destruction occurs with a further increase in pressure. It is planned that methodical provisions of the offered approach will be developed in further experiments on AE testing of objects with different physical and geometrical characteristics. The purpose of these investigations is to compile a database to be used for creation of a system of universal criterion values in regard to the object parameter and type of the AE equipment used.

Discussion of Strength Calculation and AE Testing Results

Results of strength calculation according to the finite element method performed for a pipe of D = 630 mm have shown that the destructive pressure predicted by calculation corresponds to the viscous failure area and, for the accepted material properties, is 5.9 MPa (59 kg/cm^2). The most hazardous is area 3 of the right flaw (Fig. 4), and in the area of head welding, the local flow area can occur due to bending stresses.

According to AE diagnostics data, viscous failure has begun under pressure of 50–60 kg/cm^2 in the area of the right cut and in several areas of welding of the right head. Local destruction took place in the right cut, according to data obtained from the AE method.

Thus, close agreement of the calculation data with the results of destructive tests carried out through the AE method is observed.

The suggested source concentration factor K_{AE} allowed for the rating of AE sources by hazard level and comparing data obtained by different instruments.

Conclusions

1. AE testing of steel pipe samples of D = 630 mm and D = 219 mm containing artificial defects at hydraulic testing for destruction by internal pressure with the use of two different AE systems was carried out.

2. Calculation of pipe strength with a standard size of 630×10 mm with three axial defects with the use of a two-criterion fracture diagram was performed. The calculation results have shown that destruction begins under pressure of $P = 59$ kg/cm^2.

3. According to AE testing data, defect development begins under pressure of $P = 60$ kg/cm^2.

4. It was demonstrated that application of the AE method allows for preliminary localization of destruction (at the stage before destruction) and registers the initiation of destruction with an accuracy of 5% of design pressure. Close agreement of the calculation method results with the results of the up-to-date method of technical diagnosis was obtained, which allows for tracking the physical process of destruction on-line.

5. The determination criterion of the pre-destructive state of pipe steel 17G1S according the data from AE testing was suggested. The validity of the criterion is confirmed by coincidence of the main results of its application, obtained due to the use of different types of AE equipment in the course of destructive tests on objects having different geometrical parameters.

References

1. Technical diagnostics. Acoustic-emission diagnostics. General requirements. GOST Ru 52727-2007
2. Rules for the organization and conduct of acoustic emission control of pressure vessels, boilers, apparatus and process pipelines. PB 03-593-03. Moscow (2003)
3. Patent FGUP «CNII named acad. Krylov» RU 2156456. The method for detecting defects in the welding process in the welds and for determining their location by acoustic signals, Bulletin No. 26 (2000)
4. Kovalev DN, Nefedyev EJ, Tkachev VG (2012) Acoustic emission control testing of steel corrugated pipes of circular and static loading. In: Radkevich MM, Evgrafov AN (eds) Materials of 2nd international conference modern engineering, science and education. SPb.: Publishing House of Polytechnic University, pp 382–390
5. Nefedyev EJ (2013) The use of acoustic emission method with spectral analysis of signals to determine the parameters of a leak in the pipe-line ITER. In: Radkevich MM, Evgrafov AN (eds) Modern engineering. science and education: materials of 3rd international conference. SPb.: Publishing House of Polytechnic University, pp 347–355
6. Belogur VP, Semashko NA, KuleshAA (2016) Methodological approaches to the diagnosis of products from titanium alloys by acoustic emission (AE). In: Radkevich MM, Evgrafov AN (eds) Modern engineering. science and education: materials of 5th international conference. SPb.: Publishing House of Polytechnic University, pp 350–358
7. Gomera VP, Nefedyev EJ, Smirnov AD et al (2016) Application of acoustic emission method for control of quality of welding joint in its production process. In: Radkevich MM, Evgrafov AN (eds) Modern engineering. Science and education: Materials of 5th international conference. SPb.: Publishing House of Polytechnic University, pp 376–389
8. Elchaninov GS, Nosov VV (2011) Methods of assessing the resource difficult loaded welded joints. In: Radkevich MM, Evgrafov AN (eds) Modern engineering. Science and education: materials of 1st international conference. SPb.: Publishing House of Polytechnic University, pp 212–218

9. Lakhova EN, Nosov VV (2012) Assessment of the state of critical loaded structures. In: Radkevich MM, Evgrafov AN (Eds) Modern engineering. Science and education: materials of 2nd international conference. SPb.: Publishing House of Polytechnic University, pp 445–453

10. Nefedyev E, Gomera V, Sudakov A (2014) Application of acoustic emission method for control of manual arc welding, submerged arc welding. In: Proceedings of the EWGAE-2014, Dresden, Germany, 3–5 Sept 2014 (in CD-ROM)

11. Nefedyev EJ, Gomera VP, Smirnov AD (2016) Use of the capabilities of acoustic-emission technique for diagnostics of separate heat exchanger elements. In: Evgrafov A (ed) Advances in mechanical engineering. Lecture Notes in mechanical engineering. Springer International Publishing, Switzerland, pp 183–194. https://doi.org/10.1007/978-3-319-29579-4_19

12. EPRI NP-7492. Evaluation of flaws in ferritic piping. ASME code appendix J: deformation plasticity failure assessment diagram (DPFAD)

13. Code ASME BPVC (2007) Division XI. The American Society of Mechanical Engineers

14. API 579-1/ASME FFS-1 (2007)

15. Norms for calculating the strength of equipment and pipelines of nuclear power plants PNAE G-7-002-86, Moscow, «Energoatomizdat» (1989)

Features of Calculating the Working Mechanism of an Excavator

Yuri A. Semenov and Nadezhda S. Semenova

Abstract Shovel excavators represent the most widespread type of earthmoving machine. The work of an excavator consists of periodically repeated cycles, each of which includes raising the filled bucket, swinging it to the dumping site, dumping the soil, swinging back and lowering the empty bucket. The main elements of the working mechanism of an excavator are the shovel boom, dipper stick, bucket and links enabling them to swing. The paper deals with structural and geometric analysis of the mechanism. The kinematics of the mechanism has been studied and the force analysis is presented.

Keywords Excavator · Shovel boom · Bucket · Hydrocylinders
Inverse structure · Calculation

Introduction

The work of an excavator consists of periodically repeated cycles, each of which includes raising the filled bucket, swinging it to the dumping site, dumping the soil, swinging back and lowering the empty bucket.

The main elements of the working mechanism of an excavator (Fig. 1) are the shovel boom (3), dipper stick (6), bucket (11) and the links that enable them to swing: those of the shovel boom (1) and (2), the dipper stick (4) and (5), and the bucket (7), (8), (9), and (10).

The swing mechanisms of the excavator's links mainly consist of two flexible links which together make up the input pair; structurally, the mechanisms comprise a cylinder and a piston [1–8].

Y. A. Semenov (✉) · N. S. Semenova
Peter the Great St. Petersburg Polytechnic University, Saint Petersburg, Russia
e-mail: tmm-semenov@mail.ru

N. S. Semenova
e-mail: tmm-nss@yandex.ru

© Springer International Publishing AG 2018
A. N. Evgrafov (ed.), *Advances in Mechanical Engineering*, Lecture Notes in Mechanical Engineering, https://doi.org/10.1007/978-3-319-72929-9_14

Fig. 1 The working
mechanism of an excavator

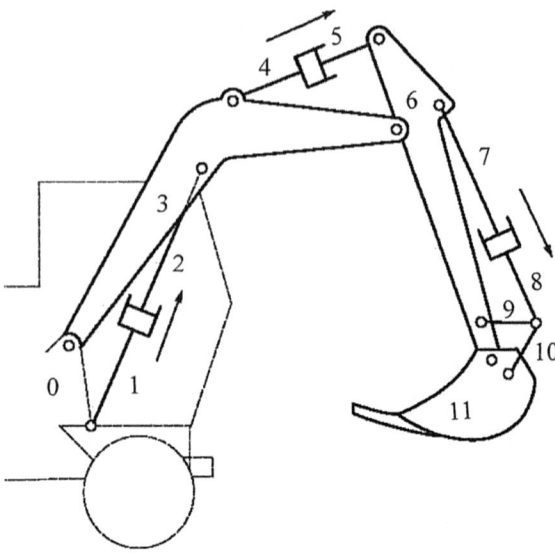

The Structural Analysis of an Excavator

Since every valid mechanism can be designed in the form of a series connection of
structural groups to a bar, it is possible to divide a mechanism into separate
structural groups. This structural decomposition of the mechanism can significantly
simplify its geometrical, kinematic and dynamic research, because structural
groups, as a rule, are described using the independent systems of the corresponding
equations of the lower degree.

    The first stage of the structural analysis of a mechanism is to produce the
structural or kinematic scheme. This presupposes counting the number N of the
flexible links and the number $P = \sum_{s=1}^{5} p_s$ of the kinematic pairs that are part of
the mechanism. Then, the mobility s of every kinematic pair is defined, as is the
total number $S = \sum_{s=1}^{5} sp_s$ of the mobilities of all of the kinematic pairs of the
mechanism.

    With the help of the received figures N, P and S for the kinematic scheme, it is
possible to identify the number w of the degrees of freedom (mobility) for the
spatial mechanism:

$$w = 6N - \sum_{s=1}^{5}(6-s)p_s = \sum_{s=1}^{5} sp_s - 6\left(\sum_{s=1}^{5} p_s - N\right) = S - 6(P-N)$$

Similarly, for a planar mechanism, we have

$$w = S - 3(P - N).$$

The working mechanism of an excavator will be considered as a planar mechanism; the spatial motion is determined only by turning the working mechanism along the vertical axis. For the planar working mechanism (see Fig. 1), we have $N = 11$, $P = S = 15$. Therefore, $w = 3$, i.e., the working mechanism, is a three-degrees-of-movement one.

The next step in the structural analysis of a mechanism is identifying its structural groups, i.e., normal kinematic chains (c) fulfilling the condition

$$n_c = w_c = S_c - 3(P_c - N_c),$$

where n_c is the number of inputs (engines) along the chain.

This problem can be solved with the help of graphs. In the graph of the mechanism, the points of the graph correspond to the links and the ribs—to the kinematic pairs. Moreover, the number of ribs linking the adjacent points is equal to the mobility of the respective kinematic pair. The thick lines in the graph show the root ribs corresponding to the input kinematic pairs.

Figure 2 shows the graph for the working mechanism of the excavator.

Let us point out the numerical feature of the subgraph, corresponding to the planar structural group: the difference between the total number S_c of the ribs of the subgraph and the number n_c of the root (thickened) ribs, which is equal to the number of the non-root (thin) ribs, is a multiple of three for a planar mechanism. The figure shows that the working mechanism of the excavator is made up of four structural groups. The formation of a mechanism can be presented by means of its structural graph, in which the points correspond to the structural groups and the ribs to the connections between the groups. The number of links and inputs of the groups are shown inside the points. Thus, the working mechanism of the excavator

Fig. 2 The graph for the working mechanism of the excavator

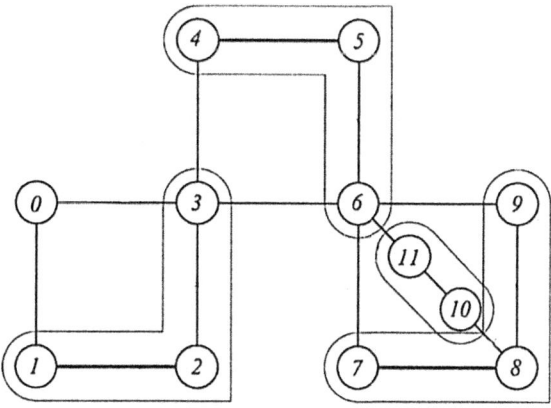

Fig. 3 The structural graph
for the working mechanism of
the excavator

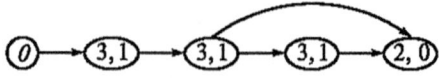

is formed by means of a series connection of three one-degree-of-movement
three-link groups and an Assur-group (Fig. 3) [9].

Geometric Analysis of the Excavator

The purpose of the geometric analysis is to define the functions of the position of
the mechanism, i.e., the dependencies of the output coordinates, characterizing the
position of its links, on the given input coordinates. The movement $q_i (i = 1, 2, 3)$ of
pistons relative to cylinders is viewed as the generalized (input) coordinates in the
three swinging mechanisms.

In order to identify the output coordinates via conventional disconnection of
certain kinematic pairs, the closed kinematic chain of the working mechanism is
modified to the open kinematic chains of the "tree" type (Fig. 4). Thus, the angles
$\varphi_1, \varphi_3, \varphi_4, \varphi_6, \varphi_7, \varphi_9, \varphi_{10}, \varphi_{11}$ characterizing the position of the links relative to
the bar will be the output coordinates.

The geometric analysis lies in identifying the angles $\varphi_1, \varphi_3, \varphi_4, \varphi_6, \varphi_7,$
$\varphi_9, \varphi_{10}, \varphi_{11}$ according to the given relative motion q_1, q_2, q_3 of the links consti-
tuting the input pairs.

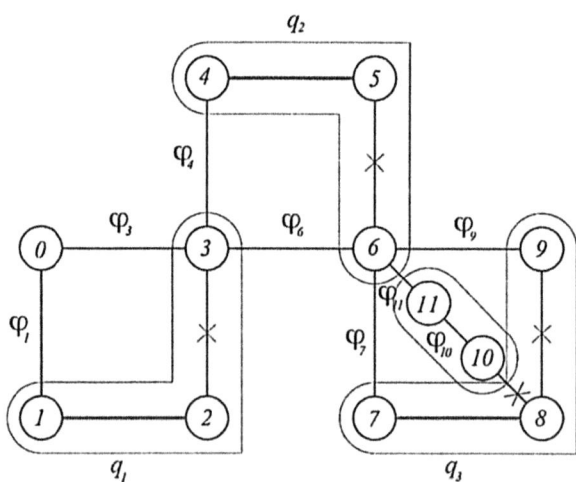

Fig. 4 The graph for the working mechanism of the excavator with geometric coordinates

Let us solve the problem of the position of the shovel boom (3) [10]. When projecting the vector equation $\overline{AC} = \overline{AD} + \overline{DC}$ on the coordinate axis (Fig. 5), we get two trigonometric equations

$$\left.\begin{array}{l} (AB_1 + q_1 + B_2C)\cos\varphi_1 = x_D - x_A + DC\cos\varphi_3, \\ (AB_1 + q_1 + B_2C)\sin\varphi_1 = y_D - y_A + DC\sin\varphi_3 \end{array}\right\} \quad (1)$$

with two unknowns φ_1 and φ_3.

Equation (1) are rewritten as

$$\left.\begin{array}{l} (AB_1 + q_1 + B_2C)\cos\varphi_1 - DC\cos\varphi_3 = x_D - x_A, \\ (AB_1 + q_1 + B_2C)\sin\varphi_1 - DC\sin\varphi_3 = y_D - y_A. \end{array}\right\} \quad (2)$$

If both parts of Eq. (2) are squared and added, it can be stated that

$$\cos(\varphi_1 - \varphi_3) = \frac{(AB_1 + q_1 + B_2C)^2 + DC^2 - (x_D - x_A)^2 - (y_D - y_A)^2}{2(AB_1 + q_1 + B_2C) \cdot DC};$$

$$\sin(\varphi_1 - \varphi_3) = \sqrt{1 - \cos^2(\varphi_1 - \varphi_3)}, \quad \text{i.e.} \varphi_{13} = \varphi_1 - \varphi_3.$$

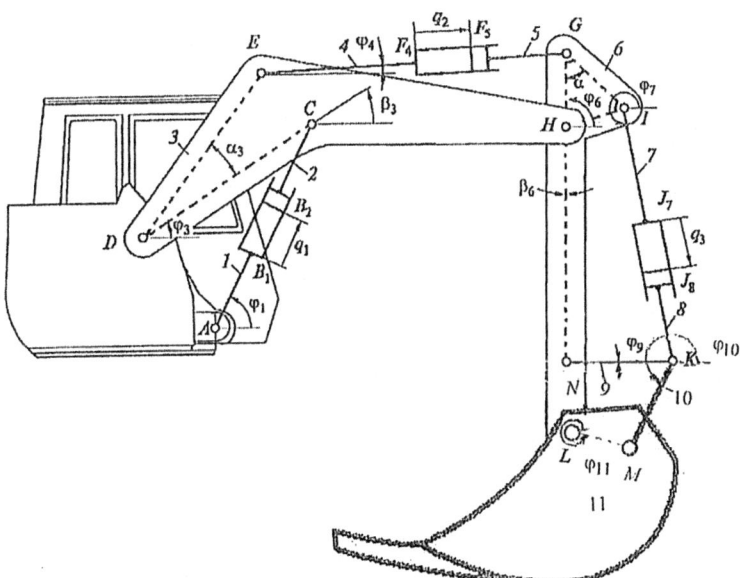

Fig. 5 The working mechanism of an excavator with geometric coordinates

The obtained relative angle φ_{13} makes it possible, on the basis of the system of equations

$$\left.\begin{array}{l}(AB_1 + q_1 + B_2C - DC \cos\varphi_{13})\cos\varphi_1 - DC \sin\varphi_{13}\sin\varphi_1 = x_D - x_A, \\ DC \sin\varphi_{13}\cos\varphi_1 + (AB_1 + q_1 + B_2C - DC \cos\varphi_{13})\sin\varphi_1 = y_D - y_A \end{array}\right\},$$

to identify, using Cramer's rule, that

$$\cos\varphi_1 = \frac{(x_D - x_A)(AB_1 + q_1 + B_2C - DC \cos\varphi_{13}) + (y_D - y_A)DC \sin\varphi_{13}}{(AB_1 + q_1 + B_2C - DC \cos\varphi_{13})^2 + (DC \sin\varphi_{13})^2};$$

$$\sin\varphi_1 = \frac{(y_D - y_A)(AB_1 + q_1 + B_2C - DC \cos\varphi_{13}) - (x_D - x_A)DC \sin\varphi_{13}}{(AB_1 + q_1 + B_2C - DC \cos\varphi_{13})^2 + (DC \sin\varphi_{13})^2},$$

i.e., angle φ_1 and angle $\varphi_3 = \varphi_1 - \varphi_{13}$.

Similar equations of the geometric analysis are written for the rest of the one-degree-of-movement three-link groups:

$$\left.\begin{array}{l}(EF_4 + q_2 + F_5G)\cos\varphi_4 - HG \cos\varphi_6 = x_H - x_E, \\ (EF_4 + q_2 + F_5G)\sin\varphi_4 - HG \sin\varphi_6 = y_H - y_E, \end{array}\right\} \tag{3}$$

$$\left.\begin{array}{l}(IJ_7 + q_3 + J_8K)\cos\varphi_7 - LK \cos\varphi_9 = x_N - x_I, \\ (IJ_7 + q_3 + J_8K)\sin\varphi_7 - LK \sin\varphi_9 = y_N - y_I, \end{array}\right\}$$

where

$$\begin{array}{ll}x_H = x_C + CH \cos(\varphi_3 - \beta_3), & x_E = x_D + DE \cos(\alpha_3 + \varphi_3), \\ y_H = y_C + CH \sin(\varphi_3 - \beta_3), & y_E = y_D + DE \sin(\alpha_3 + \varphi_3), \\ x_N = x_H - HL \cos(\varphi_6 + \beta_6), & x_I = x_G - GI \cos(\varphi_6 + \alpha_6), \\ y_N = y_H - HL \sin(\varphi_6 + \beta_6), & y_I = y_G - GI \sin(\varphi_6 + \alpha_6), \end{array} \tag{4}$$

$$\begin{array}{ll}x_C = x_A + (AB_1 + q_1 + B_2C)\cos\varphi_1, & y_C = y_A + (AB_1 + q_1 + B_2C)\sin\varphi_1, \\ x_G = x_E + (EF_4 + q_2 + F_5G)\cos\varphi_4, & y_G = y_E + (EF_4 + q_2 + F_5G)\sin\varphi_4. \end{array} \tag{5}$$

Using an algorithm similar to the above-mentioned one, let us calculate the angles of rotation for the links $\varphi_4, \varphi_6, \varphi_7, \varphi_9$.

From the geometric equations of the Assur-group [link (10) and bucket (11)]

$$\left.\begin{array}{l}LM \cos\varphi_{11} - KM \cos\varphi_{10} = x_K - x_L, \\ LMc \sin\varphi_{11} - KM \sin\varphi_{10} = y_K - y_L. \end{array}\right\}, \tag{6}$$

angles φ_{10} and φ_{11} are calculated. Here,

$$\begin{array}{ll}x_L = x_H + HL \cos(\varphi_6 + \beta_6), & x_K = x_I + (IJ_7 + q_3 + J_8K)\cos\varphi_7, \\ y_L = y_H + HL \sin(\varphi_6 + \beta_6), & y_K = y_I + (IJ_7 + q_3 + J_8K)\sin\varphi_7. \end{array}$$

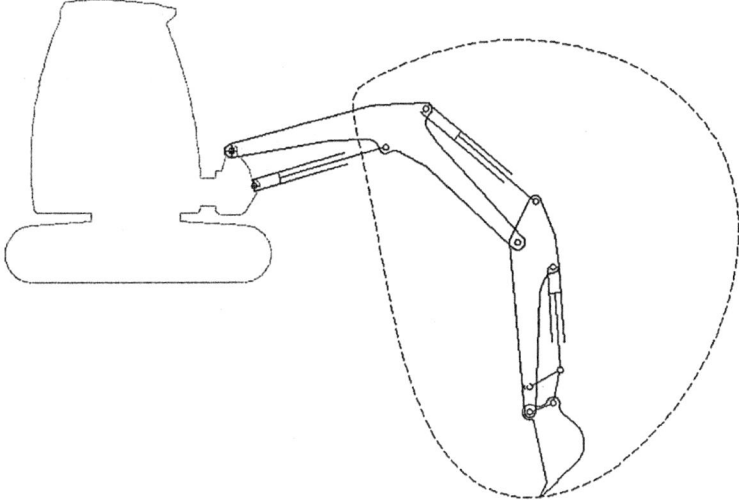

Fig. 6 The trajectory of the apex of the bucket

The Cartesian coordinates of the pole of the bucket ("back shovel") are

$$x_M = x_L + LM \cos \varphi_{11}, \quad y_M = y_L + LM \sin \varphi_{11}.$$

Figure 6 shows the trajectory of the apex of the bucket.

Selecting the Laws of the Program Motion

The laws of the program for the motion $q_i(t)$ $(i = 1, 2, 3)$ must exclude the possibility of violating the continuity of the function of the position and their first derivatives. At the same time, the three-period structure of the interval of motion turns out to be general, and the relative motion of the links is considered as the total of three zones: acceleration $(0 - t_1)$, steady motion $(t_2 - t_1)$ and run-out $(t_3 - t_2)$.

The rectangular or uniformly accelerated law of motion is adopted for the swinging mechanism of the shovel boom (Fig. 7). The changes in velocity can be presented with the help of the following equation:

$$\dot{q}_1(t) = \frac{h_1}{t_1} t\eta(t) - \frac{h_1}{t_1}(t - t_1)\eta(t - t_1) - \frac{h_1}{t_1}(t - t_2)\eta(t - t_2), \qquad (7)$$

where the unit function $\eta(t - t_i)$ is calculated.

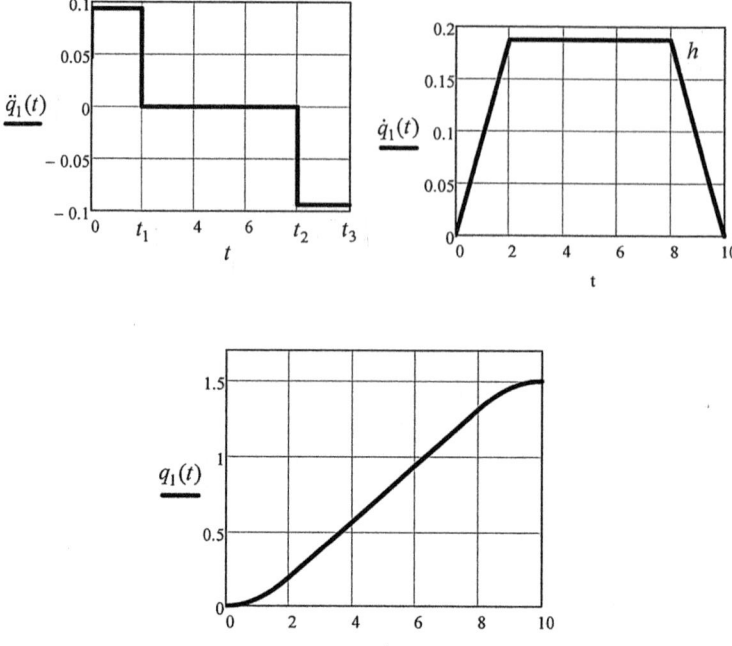

Fig. 7 The laws of the program for the motion and their derivatives

In order to identify the motion, let us integrate the function (7):

$$q_1(t) = \frac{h_1}{t_1}\frac{t^2}{2}\eta(t) - \frac{h_1}{t_1}\frac{(t-t_1)^2}{2}\eta(t-t_1) - \frac{h_1}{t_1}\frac{(t-t_2)^2}{2}\eta(t-t_2) + C.$$

From the initial condition $q_1(0) = 0 + C = 0$, let us calculate $C = 0$. Using the boundary condition $q_1(t_3) = q_{1\max}$, let us find the displacement amplitude

$$h_1 = \frac{2q_{1\max} \cdot t_1}{t_3^2 - (t_3 - t_1)^2 - (t_3 - t_2)^2}.$$

The condition of the existence of the shovel boom's swinging mechanism helps us to conclude that

$$q_{1\min} < q_1(t) < q_{1\max},$$

where

$$q_{1\min} = -(AB_1 + B_2 C) - |AD - DC|,$$

$$q_{1\max} = AD + DC - (AB_1 + B_2 C).$$

In order to calculate the velocity, let us differentiate the function according to time (7):

$$\ddot{q}_1(t) = \frac{h_1}{t_1}\eta(t) - \frac{h_1}{t_1}\eta(t - t_1) - \frac{h_1}{t_1}\eta(t - t_2).$$

Analogous laws of motion are adopted for the swinging mechanisms of the dipper stick and the bucket.

The drawback of the rectangular law is the abrupt change in the function $\ddot{q}_1(t)$ (mild blow), leading to unwanted dynamic effects. In this respect, the cosinusoidal (sinusoidal) or polynomial laws of motion are more effective.

The Inverse Problem of the Geometric Analysis

The inverse geometric problem is formulated in the following way: using the two given Cartesian coordinates of the pole of the bucket, for instance, $x_L(t)$ and $y_L(t)$, it is required that we calculate the two input coordinates $q_1(t)$ and $q_2(t)$. To do this, it is necessary to "freeze" the coordinate q_3, which means that the angles $\varphi_7, \varphi_9, \varphi_{10}, \varphi_{11}$ will not change with time.

Owing to the fact that the number of inputs ($n = 2$) coincides with the number of output coordinates ($m = 2$), it is simpler to solve the inverse problem with the help of the inverse structure of the mechanism, when the inputs and outputs swap places with each other. Let us build the graph of the inverse structure (Fig. 8). Here, thick lines show the given coordinates $x_L(t)$ and $y_L(t)$. Next, let us divide the graph of the mechanism into subgraphs corresponding to the new structural groups

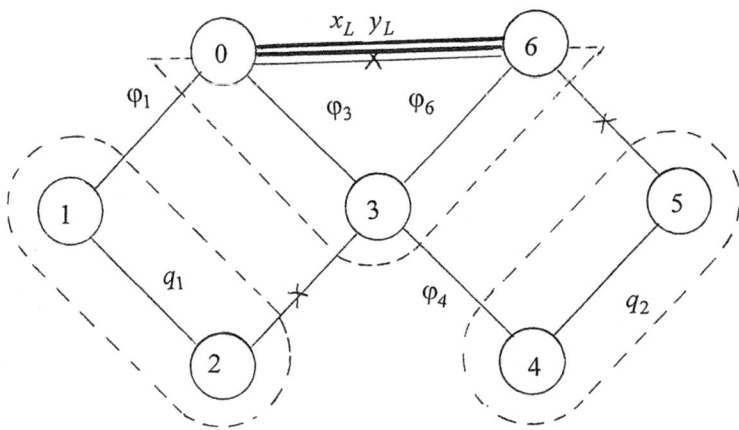

Fig. 8 The graph of the inverse structure for the working mechanism of the excavator

(the number of thin lines is multiple to three). The graph falls into three structural groups: 6–3, 1–2, 4–5.

Using the geometric equations of the structural group 6–3,

$$\left. \begin{array}{l} DH \cos \varphi_3^* + HL \cos \varphi_6 = x_L, \\ DH \sin \varphi_3^* + HL \sin \varphi_6 = y_L \end{array} \right\},$$

let us calculate the angles $\varphi_3 = \varphi_3^* + \alpha$ and φ_6 with the help of Cramer's rule, where $\alpha = \angle CDH = \arccos\left(\frac{DC^2 + HD^2 - CH^2}{2DC \cdot HD}\right)$.

The left and right parts of the geometric equations of group 1–2,

$$\left. \begin{array}{l} (AB_1 + q_1 + B_2C) \cos \varphi_1 - DC \cos \varphi_3 = x_D - x_A, \\ (AB_1 + q_1 + B_2C) \sin \varphi_1 - DC \sin \varphi_3 = y_D - y_A \end{array} \right\},$$

are squared and added, thus

$$q_1 = \sqrt{(x_D - x_A + DC \cos \varphi_3)^2 + (y_D - y_A + DC \sin \varphi_3)^2} - AB_1 - B_2C,$$

$$\cos \varphi_1 = \frac{x_D - x_A + DC \cos \varphi_3}{AB_1 + q_1 + B_2C}, \quad \sin \varphi_1 = \frac{y_D - y_A + DC \sin \varphi_3}{AB_1 + q_1 + B_2C},$$

As before, using the system of equations of group 4–5,

$$\left. \begin{array}{l} (EF_4 + q_2 + F_5G) \cos \varphi_4 - HG \cos \varphi_6 = x_H - x_E, \\ (EF_4 + q_2 + F_5G) \sin \varphi_4 - HG \sin \varphi_6 = y_H - y_E \end{array} \right\},$$

let us find

$$q_2 = \sqrt{(x_H - x_E + HG \cos \varphi_6)^2 + (y_H - y_E + HG \sin \varphi_6)^2} - EF_4 - F_5G,$$

$$\cos \varphi_4 = \frac{x_H - x_E + HG \cos \varphi_6}{EF_4 + q_2 + F_5G}, \quad \sin \varphi_4 = \frac{y_H - y_E + HG \sin \varphi_6}{EF_4 + q_2 + F_5G}.$$

The Kinematic Analysis of the Excavator

The purpose of the kinematic analysis is to define velocities and accelerations of the links of the excavator. Let us differentiate the group equations according to time (2):

$$\left. \begin{array}{l} (AB_1 + q_1 + B_2C) \sin \varphi_1 \dot{\varphi}_1 - DC \sin \varphi_3 \dot{\varphi}_3 = \dot{q}_1 \cos \varphi_1, \\ (AB_1 + q_1 + B_2C) \cos \varphi_1 \dot{\varphi}_1 + DC \cos \varphi_3 \dot{\varphi}_3 = \dot{q}_1 \sin \varphi_1, \end{array} \right\} \tag{8}$$

whence we find

$$\dot\varphi_1 = \frac{\dot q_1}{(AB_1 + q_1 + B_2C)\mathrm{tg}(\varphi_1 - \varphi_3)}; \quad \dot\varphi_3 = \frac{\dot q_1}{DC\,\sin(\varphi_1 - \varphi_3)}.$$

Differentiating (8) according to time, we obtain a system of linear equations

$$\left.\begin{array}{l} (AB_1 + q_1 + B_2C)\sin\varphi_1\ddot\varphi_1 - DC\,\sin\varphi_3\ddot\varphi_3 = M_1, \\ -(AB_1 + q_1 + B_2C)\cos\varphi_1\ddot\varphi_1 + DC\,\cos\varphi_3\ddot\varphi_3 = M_2, \end{array}\right\}$$

where

$$M_1 = -2\dot q_1\dot\varphi_1\sin\varphi_1 - (AB_1 + q_1 + B_2C)\cos\varphi_1\dot\varphi_1^2 + DC\,\cos\varphi_3\dot\varphi_3^2 + \ddot q_1\cos\varphi_1,$$
$$M_2 = 2\dot q_1\dot\varphi_1\cos\varphi_1 - (AB_1 + q_1 + B_2C)\sin\varphi_1\dot\varphi_1^2 + DC\,\sin\varphi_3\dot\varphi_3^2 + \ddot q_1\sin\varphi_1,$$

whence we find

$$\ddot\varphi_1 = \frac{M_1\cos\varphi_3 + M_2\sin\varphi_3}{(AB_1 + q_1 + B_2C)\sin(\varphi_1 - \varphi_3)}; \quad \ddot\varphi_3 = \frac{M_2\sin\varphi_1 + M_1\sin\varphi_1}{DC\,\sin(\varphi_1 - \varphi_3)}.$$

The equations of the geometric analysis are differentiated in the same way for other structural groups. Using the given dimensions of the links of the excavator and the signs of the velocities of input pairs, it is fairly easy to define the maximum depth of digging and the maximum dumping height of the bucket.

The Force Analysis of the Excavator

Identification of the driving forces and reactions in kinematic pairs is performed with the help of kinetostatic equations in the direction opposite to that of the formation of the excavator. The kinetostatic equations for links (10) and (11) (Fig. 9) are presented as follows:

$$R_{6,11x} + R_{9,10x} + \Phi_{11x} + P_x = 0;$$
$$R_{6,11y} + R_{9,10y} + \Phi_{11y} - G_{11} + P_y = 0;$$
$$-R_{6,11y}(x_M - x_L) - R_{6,11x}(y_L - y_M) - (\Phi_{11y} - G_{11} + P_y)(x_M - x_{C_{11}})$$
$$+ (\Phi_{11x} + P_x)(y_M - y_{C_{11}}) + M_{C_{11}z}^{(\Phi)} = 0; R_{9,10y}(x_K - x_M) - R_{9,10x}(y_K - y_M) = 0.$$

where R_{ij} is the constraining force of an i-link to j-link, Φ_i is the force of inertia, G_i is gravitation, and P_x, P_y are the components of the working load modified to the center of mass of the bucket.

Hydraulic shovel excavators, equipped with a back shovel, are used to dig the soil: by swinging the dipper stick (with the shovel boom staying motionless)

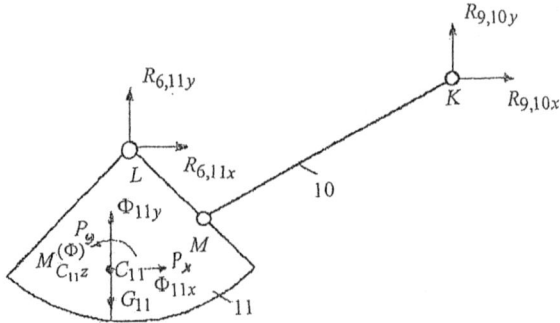

Fig. 9 Assur-group (links 10 and 11) with the applied forces

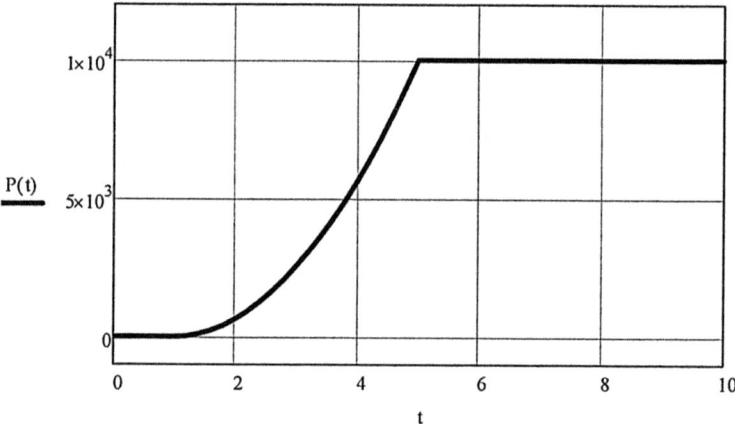

Fig. 10 The law of change efforts on the cutting edge of the bucket

or swinging the bucket (along with the motionless shovel boom and the dipper stick) or using a mixed technique.

The effort applied to the cutting edge of the bucket changes according to the law shown in Fig. 10.

The kinetostatic equations for links (7), (8), and (9) (Fig. 11) are

$$R_{67x} + R_{69x} + R_{10,9x} = 0;$$
$$R_{67y} + R_{69y} + R_{10,9y} = 0;$$
$$-R_{69x}(y_N - y_K) + R_{69y}(x_K - x_N) = 0;$$
$$-R_{67x}(y_I - y_K) + R_{67y}(x_I - x_K) = 0.$$

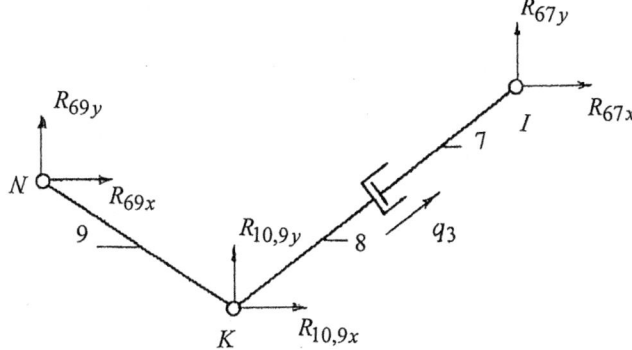

Fig. 11 Three one-degree-of-movement three-link group with the applied forces

From the equilibrium equations of links (7) and (8),

$$R_{67x} + R_{10,9x} + R_{98x} = 0,$$
$$R_{67y} + R_{10,9y} + R_{98y} = 0,$$

let us define the reactions R_{98x}, R_{98y}.

The equilibrium equation of link (8),

$$(R_{98x} + R_{10,9x})\cos(\varphi_7 + \pi) + (R_{98y} + R_{10,9y})\sin(\varphi_7 + \pi) = Q_3$$

enables us to calculate the propulsive force Q_3 acting on cylinder (7) on the part of piston (8).

The propulsive forces Q_2 and Q_1 are defined in the same way. The required hydraulic engines are selected on the basis of the received laws of the variation of propulsion forces.

To improve the calculation data performed in the "Excel", "Mathcad" and "MatLab" programs, we have proposed software applications that allow for creating a computer animation of the kinematic scheme of the mechanism [11].

References

1. Yahya HZ, Lakmal DS, Kaspar A (2003) A generalized Newton method for identification of closed-chain excavator arm parameters. In: Proceedings of the 2003 IEEE International Conferenee on Robotics & AutomationTaipei, Taiwan, 14–19 Sept, 1003 pp
2. Chang Lv, Zhang J (2011) Excavating force analysis and calculation of dipperhandle. IEEE, pp 124–130
3. Srushti HB, Ravi Prakash N, Jadeja SB (2013) Modelling of robotic manipulatorARM. Int J Mech Eng Technol (IJMET) 4(3), 125–129. ISSN Print: 0976-6340, ISSN Online: 0976-6359
4. Ghosh A, Mallik AK (1988) Theory of mechanisms and machines, 2nd ed. Affiliated East-West Press Private Limited, New Delhi, India, 586 p

5. Isakson AA (1975) Universal Shovel hydraulic excavator EO-5122. Construction and road machines, no. 2 (with 7–8 (rus))
6. Dotsenko AI, Karasev GN, Kustarev VG, Shestopalov KK (2012) Machines for earthworks: a textbook for University students. BASTET, Moscow, 688 p (rus)
7. Berkman L (1993) To hydraulic excavators. Higher. wk., Moscow, 371 p (rus)
8. Krutikov KE (1994) Excavators. Mashinostroenie, Moscow, 391 p (rus)
9. Kolovsky MZ, Evgrafov AN, Semenov YuA, Slousch AV (2000) Advanced theory of mechanisms and machines. Springer, Berlin Heidelberg, New York, 394 p
10. Semenova NS, Semenov YuA (2009) Course project "Investigation of lifting-transport and building-road cars". Theory Mech Mach 7(2):61–71 (rus)
11. Evgrafov AN, Petrov GN (2008) Computer animation of a kinematic schemes in EXCEL and MATHCAD programs. Theory Mech Mach 6(11):71–80 (rus)

Localization of Plastic Deformation HCP—Crystals During Indentation and Scratching

Margarita A. Skotnikova, Galina V. Ivanova, Alexander A. Popov and Olga V. Paitova

Abstract The paper examines localization of plastic deformation HCP crystals during indentation and scratching. For example, the anisotropic titanium theoretically and experimentally that the profiling roller (bulk) of the imprint or the risks generated in the process of measuring the hardness or scratching depends on the ability of the material to transverse plastic deformation (with uniform strain hardening). It is shown that the smaller the capacity, the sooner the localization of plastic deformation in a narrow local region along the edge of the concentration, and the imprint or impress in the metal is precisely the same shape of the indenter or cutter.

Keywords Titanium · Crystallography · Indentation · Scratching
Imprint · Impress · Transverse and longitudinal plastic deformation

Introduction

Modern advances in the fields of physics and mechanics of the contact interaction provide the opportunity to consider the process of the plastic deformation, destruction and energy release, as well as the physic-chemical processes, discretely (step-like). This all coexisted with the changes of the retension waves [1–4]. Development of the considerations concerning discrete features of the process of the plastic deformation leads to the necessity to localize the deformation

M. A. Skotnikova (✉) · G. V. Ivanova · A. A. Popov · O. V. Paitova
Peter the Great Saint-Petersburg Polytechnic University, Saint-Petersburg, Russia
e-mail: skotnikova@mail.ru

G. V. Ivanova
e-mail: galura@yandex.ru

A. A. Popov
e-mail: alexandr-popov92@mail.ru

O. V. Paitova
e-mail: olja.stern@gmail.com

© Springer International Publishing AG 2018   143
A. N. Evgrafov (ed.), *Advances in Mechanical Engineering*, Lecture Notes
in Mechanical Engineering, https://doi.org/10.1007/978-3-319-72929-9_15

representing the way of the express delivery of stress concentrations by the sample. Relaxation of one stress raiser has to generate emergence in other point of a sample of the new stress raiser, and this process must relay to spread on the sample resulting in local structural and phase transformations of deformable crystal, which in General remains structure-stable. Therefore, the deformed crystal as a dissipative system is widely uses different forms of localization of deformation at all structural levels: from the formation of macro- neck grain boundary sliding to the localization of deformation along slip bands of dislocations, twins, and surfaces of the concentration.

The purpose of this work was to study the regularities of elastoplastic deformation of the material surface of the workpiece in the process of indentation by the indenter or scratching with a cutter, to establish the nature of localization (concentration) of plastic deformation on the basis of the analysis of the shape of the imprint of the HCP crystals in the contact zone with the tool.

Methods and Materials

Material for the research was the billet of titanium alloy PT-3V (Ti-4, 2 Al-1, 6 V) in the heat-treated (quenched) and annealed state.

By method of color electrochemical coloring on the surface of sample's surface detected the HCC—grains, the hexagonal axes [0001] of which were oriented perpendicularly and parallel to the surface and were painted, respectively, in yellow and lilac colors.

For indenting the surface of the material and analyzing the shape of the prints, the PMT-3M instrument and FUTURE-TECH automated microhardness (Japan) were used; optical microscopes MMP-4; MBS-9; μVizo®-MET micro-vision; IM7200 MEIJI TECHNO (Japan) with Thixomet image analyzer. Surface profile control after plastic deformation was carried out using the German profilometer Mahr Surf PS1. The length of the tracing was 4.0 mm.

Results of the Research

Features of elastic-plastic deformation of near-surface volume of the workpiece materials during indentation of the indenter or scratched with a cutter and its localization in the area of actual contact with the surface of the concentrator have not yet been fully studied, especially in titanium and its alloys with an anisotropic HCC crystal lattice. It is known that their elasticity modules (E_{hkl}) and shear (G_{hkl}) are directly related to the forces of interatomic interaction, which are distinctive in different crystallographic directions and are a measure of resistance to deformation: volume, transverse and longitudinal [4].

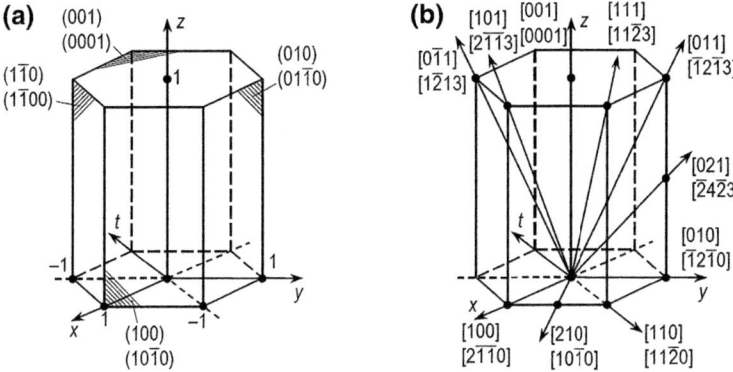

Fig. 1 Three digit Miller's indexes and for digit Miller-Brave for some planes (**a**) and directions (**b**) in hexagonal lattice

In the case of indentation and scratching, the loading method is soft, as the increase of shear stresses in the metal occurs faster than normal. This increases the proportion of transverse deformation (change in the transverse dimensions of the body), which is relied upon in the Poisson's ratios (μ) and compliance (S_{12}). For most metals and alloys with a cubic crystallographic lattice, the ratio of the variation of the transverse to the longitudinal dimensions of the body (μ) is in the narrow range 0.25 ... 0.35 [1]. For titanium μ can vary significantly from 0.322 to 0.485 depending on the crystallographic orientation of the applied stresses.

Figure 1 shows the appearance of an HCC lattice with Miller's indexes crystallographic indices of planes (a) and directions (b).

The calculation of the total volumetric deformation (ΔV) taking into account the anisotropy of the HCC crystal lattice was carried out according to the formula (1).

$$\Delta V = 3(1 - 2\mu)/E \cdot \sigma_o \qquad (1)$$

The results of the calculation are presented in Table 1.

Table 1 Influence crystallographic orientation penetration indenter in HCC lattice titanium on volume strain in that directions

Crystallo-graphic orientation	E_{hkl}, GPa, modulus of elasticity	μ, Poisson's ratio, for transverse deformation	G_{hkl}, GPa, shear modulus	$S_{12} \times 10^{12}$, cm²/length compliance ratio	$3(1-2\mu)/E \times 10^3$, GPa^{-1} volume deformation
[0001]	143.26	0.322	54.18	−0.2248	7.4549
[1010]	104.38	0.443	36.18	−0.4244	3.2764
[1012]	97.15	0.476	32.92	−0.4900	1.4822
[1011]	95.23	0.485	32.06	−0.5093	0.9451

Estimation of the Fraction of Transverse and Total Volume Deformation at Loading of HCP-Grains Along the Hexagonal Axis [0001]

It is known that the roller profile around an imprint or impress which is formed in the course of measurement of hardness or a scratching depends on ability of material to transverse plastic deformation (to uniform deformation hardening: $\theta = \Delta\sigma/\Delta\varepsilon$).

It is possible to believe that with reduction of size of transverse plastic deformation and uniform deformation hardening, localization of plastic deformation in narrow local area along an edge of the concentrator will occur earlier, and the imprint or impress in metal will more precisely repeat a form of the indenter or cutter.

As the calculation results showed, loading of HCP-grains along the hexagonal axis [0001] leads to the achievement of maximum volume deformation with a minimum fraction of the transverse deformation (Table 1).

If the compressive stresses are oriented along the crystallographic direction [0001] of the HCP-lattice with strong interatomic bond and do not coincide with any of the sliding systems, then along with the large volume deformation (ΔV), the proportion of transverse deformation remains small (μ) (Table 1). There is no significant deformation hardening in the HCP-grain in this direction, and zones of constrained deformation occur in a narrow local the area along the edge of the indenter.

As the results of the imprints analysis have shown, in this case, the roll profile around the impression in the metal increases in height and decrease in width, that is, precisely repeats the shape of the indenter.

Estimation of the Fraction of Transverse and Total Volume Deformation Under Loading of HCP-Grains Perpendicular to the Hexagonal Axis [0001]

The loading of HCP-grains perpendicular to the hexagonal axis [0001] leads to a significant decrease in the total volume of deformation and to a noticeable increase in the fraction of transverse deformation (Table 1).

If the compressive stresses are oriented perpendicular to the direction of the hexagonal axis [0001] with weaker interatomic bonds and coincide with several sliding systems, then in the contact area near the indenter, a uniform strain hardening of the metal develops, and deformation localization does not occur. The profile of the imprint roller decreases.

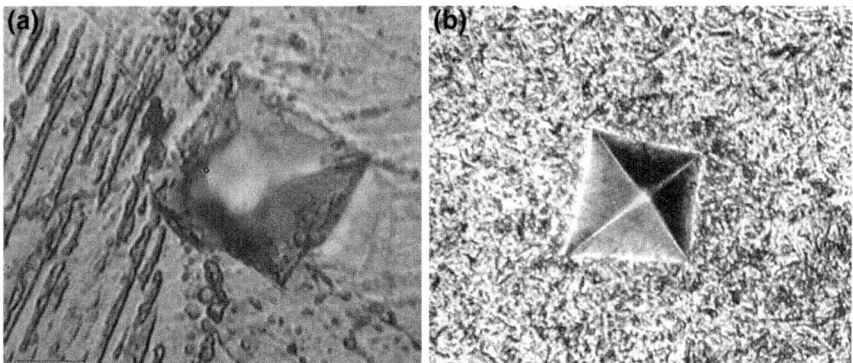

Fig. 2 Prints of hardness in titanium, with a load of 500 g in the annealed (**a**) and quenched (**b**) titanium alloy, x 400

Estimation of the Proportion of Transverse Deformation and Form of Imprints by Loading HCP-Grains in the Quenched and Annealed State

Knowing that heat treatment of a material affects the particularity of elasto-plastic deformation of near-surface volumes of workpiece materials during indentation of the indenter or scratched with a cutter.

As can be seen from Fig. 2, in comparison with the annealed state in Fig. 2a, after hardening or quenching (Fig. 2b), the imprint in the metal more precisely repeats the form of the indenter.

It can be assumed that during testing the hardness of quenched martensite of the titanium present therein oriented residual compressive stress, inhibits the uniform deformation hardening, increased localization of plastic deformation and led to the formation of prints having exactly the same shape of the indenter, (Fig. 2b).

Construction of a Diagram of the Types of Contact Interaction of Materials Depending on the Configuration of the Indenter and the Value of the Contact Stress

In this paper, we studied the conditions of longitudinal and transverse displacement of the metal, accompanying micro-indentation by an indenter or scratching by a tool, depending on the applied contact stresses, tool configuration and strength material. The diagram is constructed of the types of contact interaction of materials according to the tool configuration and magnitude of the applied contact shear stresses (Fig. 3).

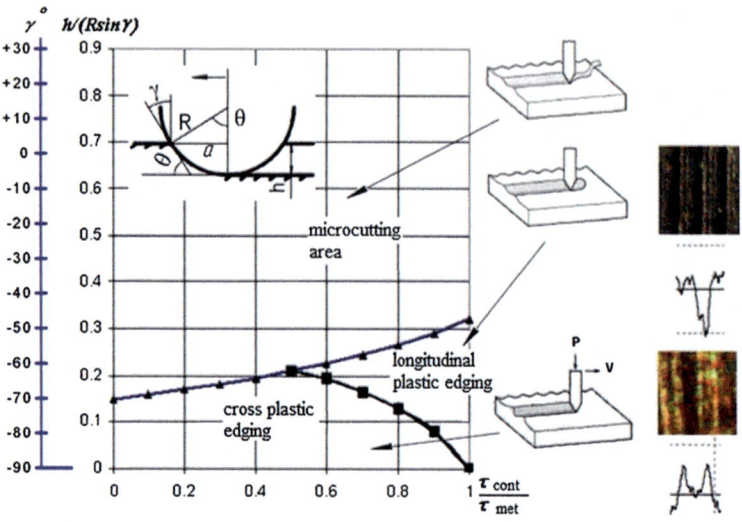

Fig. 3 Diagram of the types of contact interaction of materials depending to the tool configuration and magnitude of the applied contact shear stresses

As a parameter characterizing the configuration of the indenter was selected ratio of the penetration depth of the indenter **h** to the rounded radius of the indenter **R** and sine of face angle γ, that is, the parameter **(h/R sin γ)**, (Fig. 3). The actual contact shear stresses of the material being processed were calculated according to the relationship

$$\tau_{cont.} = \alpha G b \rho^{1/2}, \tag{2}$$

Where, α—constant approximately equal to 0.3 … 0.5, G—shear modulus, b—Burgers vector, ρ—dislocation density.

As the relative contact shear stress, the ratio of the actual contact shear stress to the ultimate resistance of the workpiece material to shear (τ_{met}) was taken, that is, the parameter ($\tau_{cont.}/\tau_{met.}$) (Fig. 3).

The appearance of piles on the edges of the imprint or impress that are produced during indentation or scratching, testified about the course of transverse (cross) plastic deformation to the direction of movement of the tool edge (Fig. 3).

The absence of piles at the edges of the imprint or impress, compaction, the formation of a roller of material in front of the moving edge of the instrument indicated the occurrence of longitudinal plastic deformation (Fig. 3).

It is shown that transverse and longitudinal plastic displacement is observed at relatively low and high contact stresses, respectively.

Analysis of surfaces obtained after machining, blanks using optical, electron microscopy and profilometry, revealed 2 types of grooves.

At relatively low contact stresses ($\tau_{cont.} \leq 0.5\tau_{met.}$)—grooves were formed with piles—as a result of the transverse plastic displacement of the metal relative to the direction of shear.

At relatively high contact stresses ($\tau_{cont.} \geq 0.5\tau_{met.}$)—formations of grooves without piles were formed as a result of longitudinal plastic displacement of the metal relative to the direction of shear.

At the face angles and contact stresses above some critical values, the transverse or longitudinal plastic displacement was replaced by microcutting and the formation of a micro-chip (Fig. 3).

Thus, it was shown that plastic edging accompanying indentation or scratching cutter into contact zone of friction pair "tool-workpiece" according to the elastically-stressed state, may be caused by plastic deformation of the longitudinal or transverse and, respectively, leading to the formation of longitudinal or transverse piles in front or from the sides with respect to the direction of movement of the tool edge.

Conclusion

With a soft method of loading (indenting or scratching), the shape of the imprint in HCP crystals depends on the anisotropy of the elastoplastic characteristics and the intensity of localization (concentration) of plastic deformation in the contact zone with the tool.

In the process of development of uniform plastic deformation (deformation hardening) by the mechanism of multistage evolution of the substructure, there are counter stress concentrators in the body of grains that hamper the localization of plastic deformation [5–10]. At the same time, the profiling roller of the imprint or the impress that result from indentation or scratching is very different.

The less the uniform deformation hardening develops in the material loaded with an indenter, the earlier the localization of plastic deformation occurs in a narrow local area along the edge of the concentrator, and the imprint or risk in the metal more precisely repeats the shape of the indenter or cutter.

References

1. Panin VE, Grinyaev YV, Danilov VI et al (1990) Structural levels of plastic deformation and fracture. Nauka, Novosibirsk, 255 pp
2. Likhachev VA, Panin VE, Zasimchuk EE et al (1989) Cooperative processes and localization of deformation. Sciences, Kiev, Dumka, 320 pp
3. Finkel VM (1970) Physics of fracture. Metallurgy, 322 pp
4. Kachanov LM (1974) Fundamentals of mechanics of fracture. Nauka, 311pp
5. Mikljaev PG, Neshpor GS, Kudryashov VG (1979) Kinetics of fracture. Metallurgy, 279 pp
6. Komanduri R (1981) New observations on the mechanism of chip formation when machining titanium alloys. Wear 69:179–188

7. Skotnikova MA, Tsvetkova GV, Lanina AA, Krylov NA, Ivanova GV (2015) Structural and phase transformation in material of blades of steam turbines from titanium alloy after technological treatment. Book lecture notes in mechanical engineering. pp 93–101
8. Skotnikova MA, Krylov NA, Ivanov EK, Tsvetkova GV (2016) Structural and phase transformation in material of steam turbines blades after highspeed mechanical effect. Lecture notes in mechanical engineering. pp 159–168
9. Skotnikova MA, Tsvetkova GV, Krylov NA, Ivanov EK (2016) Features of wear of abrasive grains depending on microcuttings speed of steels. Key Eng Mater 674:358–364
10. Skotnikova MA, Krylov NA (2017) About the nature of dissipative processes in cutting treatments of titanium vanes. Advances in mechanical engineering. Lecture notes in mechanical engineering. Selected Contributions from the Conference "Modern Engineering: Science and Education", Saint Petersburg, Russia, June 20–21, 2016. Springer—Verlag. Berlin-Heidelberg. pp 115–124

Shock Response Spectra as a Result of Linear Interactions

Valerii Tereshin

Abstract This paper provides a simplified assessment of shock response spectrum resulting from linear interactions with a table that accommodates a test object. The shock response spectrum of functions is commonly perceived as the maximum absolute acceleration of an oscillator when the base is moved according to the law of this function. The spectrum is calculated as a function of the free motion frequency of the oscillator. Its calculation warrants a solution to the problem of the global maximum of a rapidly fluctuating non-monotonic function over a long period of time. The synthesis of test bench parameters for the replication of the required shock pulse constitutes an even more complex issue. The article considers discrete and continuous linear physical models of collision processes and assesses their shock response spectra as simple analytical dependencies on the parameters of interacting bodies. The error of approximation proved quite small, due to the traditional depiction of shock response spectra in a logarithmic scale.

Keywords Shock response spectrum · Test bench · Acceleration pulse

Problem

Equipment testing for shock impact [1, 2] has been performed in various industries [3–8] for many decades. Figure 1 shows the typical form of acceleration simulated on a shock stand [9].

Standard mathematical treatment [10] of this record yielded the shock spectrum shown in Fig. 2 as a solid curve.

The angled solid line shows the acceptable level of impact, whereas the dotted line and solid line with markers show piecewise-linear approximation, its calculation described in detail below.

V. Tereshin (✉)
Peter the Great St. Polytechnic University, Saint Petersburg, Russia
e-mail: terva@mail.ru

© Springer International Publishing AG 2018
A. N. Evgrafov (ed.), *Advances in Mechanical Engineering*, Lecture Notes in Mechanical Engineering, https://doi.org/10.1007/978-3-319-72929-9_16

Fig. 1 Dependence between
shock acceleration (g) and
time (t, ms)

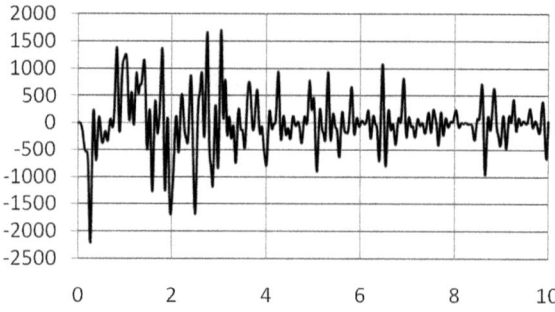

Fig. 2 Dependence between
shock acceleration (g) and
frequency (Hz)

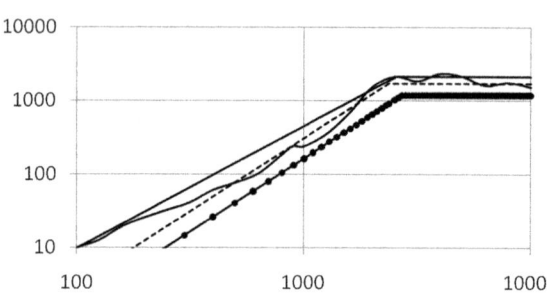

Fig. 3 Single-stage linear
elastic oscillator

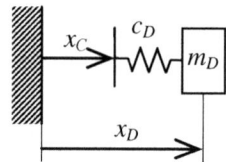

When forming the physical model of shock interaction, we assume a minor influence of the test item on table movement and negligible power dissipation at the beginning of movement when maximum accelerations of the oscillator are observed. Nonlinear characteristics of pads and other elements can have a major effect on the shock spectrum type [11], but consideration of these critical issues is outside the scope of this article. Figure 3 shows the computational scheme for a single-stage linear elastic oscillator on the moving base, where m_D, c_D and x_D stand for the mass, stiffness and absolute coordinate of the oscillator, respectively, and x_c signifies the absolute coordinate of the table. Let us consider the equation of motion for mass m_D

$$m_D\ddot{x}_D = c_D\left(x_{\hat{C}} - x_D\right) \tag{1}$$

and rewrite it in the Laplace transform p with zero initial conditions [12]

$$X_D = \frac{\omega^2}{p^2 + \omega^2} \cdot X_{\tilde{C}}, \tag{2}$$

where $\omega^2 = c_D/m_D$ is the square of free oscillation frequency of the oscillator. In order to obtain the function of table movement, let us consider the discrete and continuum models of shock interaction of stand elements [13, 14].

Discrete Model

Figure 4 shows the dynamic stand model, where m_C, c_C, x_C, m_S, c_S, x_S stand for the mass, stiffness and absolute coordinate of the table and the projectile, respectively. The equations of this system's movement are as follows:

$$\begin{cases} m_C\ddot{x}_C = -c_C x_C + c_S(x_S - x_C), \\ m_S\ddot{x}_S = -c_S(x_S - x_C). \end{cases} \tag{3}$$

Let us rewrite (3) in the Laplace transform with zero initial conditions except for the speed of the projectile, which is equal to $-v$:

$$\dot{x}_S(0) = -v, \tag{4}$$

$$\begin{cases} (m_C p^2 + c_C + c_S)X_C - c_S X_S = 0, \\ m_S(p^2 X_S + v) + c_S(X_S - X_C) = 0. \end{cases} \tag{5}$$

Then, let us divide the first equation in (5) by m_C, the second by m_S and introduce the nomenclature for squares of partial frequencies

$$\frac{c_{\tilde{C}}}{m_{\tilde{C}}} = p_1^2; \quad \frac{c_S}{m_S} = p_2^2; \quad \frac{c_S}{m_{\tilde{C}}} = p_3^2 \tag{6}$$

Fig. 4 Two-mass model of the shock stand

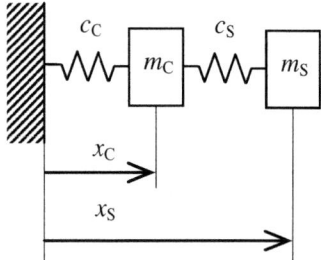

and rewrite it in matrix form

$$\begin{bmatrix} p^2 + p_1^2 + p_3^2 & -p_3^2 \\ p_2^2 & p^2 + p_2^2 \end{bmatrix} \begin{bmatrix} X_C \\ X_S \end{bmatrix} = \begin{bmatrix} 0 \\ -v \end{bmatrix}. \tag{7}$$

Let us solve the equation with respect to table movement:

$$X_C = \frac{-vp_3^2}{\Delta}. \tag{8}$$

The determinant of the equation matrix (7)

$$\Delta(p) = p^4 + (p_1^2 + p_2^2 + p_3^2)p^2 + p_1^2 p_2^2 \tag{9}$$

has four purely imaginary roots $\pm ik_{1,2}$. The finding of the root can be facilitated by changing variables $p^2 = -k^2$ and shifting from the characteristic equation $\Delta(p) = 0$ to the frequency equation

$$k^4 - (p_1^2 + p_2^2 + p_3^2)k^2 + p_1^2 p_2^2 = 0. \tag{10}$$

The frequency equation has two positive roots:

$$k_{1,2} = \sqrt{\frac{p_1^2 + p_2^2 + p_3^2}{2} \pm \sqrt{\frac{(p_1^2 + p_2^2 + p_3^2)^2}{4} - p_1^2 p_2^2}}. \tag{11}$$

To take the inverse Laplace transform of (8), let us decompose it into simple fractions:

$$\frac{-vp_3^2}{\Delta} = \frac{-vp_3^2}{(p^2 + k_1^2)(p^2 + k_2^2)} = \frac{vp_3^2}{k_1^2 - k_2^2} \left(\frac{1}{p^2 + k_1^2} - \frac{1}{p^2 + k_2^2} \right). \tag{12}$$

Thus, the Laplace transform of the table movement function for a discrete two-mass system appears as follows:

$$X_C = \frac{vp_3^2}{k_1^2 - k_2^2} \left(\frac{1}{p^2 + k_1^2} - \frac{1}{p^2 + k_2^2} \right). \tag{13}$$

Its original form with the given data analogous to the example shown in Fig. 1 was differentiated twice in time and is shown in Fig. 5.

At first glance, it seems that Figs. 1 and 5 have very little in common. But if we ignore frequencies above 10 kHz, and time after three milliseconds, we can see the

Fig. 5 Dependence between shock acceleration (g) and time (t, ms) in two-mass system stand modeling

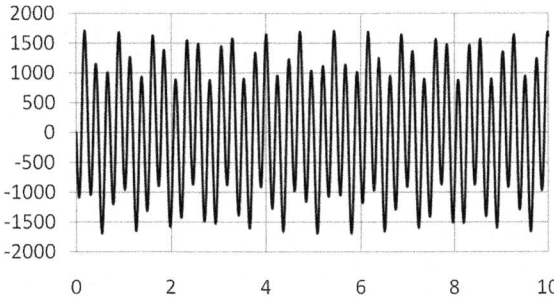

similarity in peak acceleration values and their quantities. In order to obtain the shock spectrum, let us input (13) into (2) and then decompose it into partial fractions:

$$
\begin{aligned}
X_D &= \frac{\omega^2}{p^2 + \omega^2} \cdot \frac{vp_3^2}{k_1^2 - k_2^2} \left(\frac{1}{p^2 + k_1^2} - \frac{1}{p^2 + k_2^2} \right) \\
&= \frac{vp_3^2 \omega^2}{k_1^2 - k_2^2} \left[\left(\frac{1}{k_1^2 - \omega^2} + \frac{1}{k_2^2 - \omega^2} \right) \cdot \right. \\
&\quad \left. \frac{1}{p^2 + \omega^2} - \frac{1}{k_1^2 - \omega^2} \cdot \frac{1}{p^2 + k_1^2} - \frac{1}{k_2^2 - \omega^2} \cdot \frac{1}{p^2 + k_2^2} \right].
\end{aligned}
\tag{14}
$$

Let us record the inverse Laplace transform from (14):

$$
\begin{aligned}
x_D &= \frac{vp_3^2 \omega^2}{k_1^2 - k_2^2} \left[\left(\frac{1}{k_1^2 - \omega^2} + \frac{1}{k_2^2 - \omega^2} \right) \frac{1}{\omega} \sin(\omega t) \right. \\
&\quad \left. - \frac{1}{\left(k_1^2 - \omega^2 \right) k_1} \sin(k_1 t) - \frac{1}{\left(k_2^2 - \omega^2 \right) k_2} \sin(k_2 t) \right].
\end{aligned}
\tag{15}
$$

It is important to bear in mind that the stiffness of the c_C and c_S is created by gaskets that only work in compression and cannot cause negative reactions. The time of separation can be determined as part of the set task, but we believe that maximum acceleration will be achieved before such time. The shock spectrum signifies the maximum modulus of absolute acceleration of the oscillator, so let us determine it from (15) after taking the second derivative. In general, k_1, k_2 and ω frequencies are simple and can vary significantly. Therefore, in order to evaluate maximum acceleration, let us set sine values to 1 with appropriate signs. If we denote $w_D(\omega) = \max_t |\ddot{x}_D|$, then

$$
w_D(\omega) = \frac{vp_3^2 \omega^2}{\left| k_1^2 - k_2^2 \right|} \left[\left| \frac{\omega}{k_1^2 - \omega^2} + \frac{\omega}{k_2^2 - \omega^2} \right| + \left| \frac{k_1}{k_1^2 - \omega^2} \right| + \left| \frac{k_2}{k_2^2 - \omega^2} \right| \right].
\tag{16}
$$

After transforming the assessment to the logarithmic scale, we can simplify it even further and reduce it to a piecewise linear function. So, for low frequencies, ω can be set at

$$w_{D1}(\omega) = \frac{vp_3^2\omega^2}{|k_1 - k_2|k_1 k_2}, \tag{17}$$

and for high frequencies,

$$w_{D2}(\omega) = \frac{vp_3^2}{|k_1^2 - k_2^2|}(2\omega + k_1 + k_2). \tag{18}$$

In the latter congruence, we should not neglect the aggregate of $k_1 + k_2$ if $2\omega \gg k_1 + k_2$, since with large numerical values of $k_1 + k_2$, logariphica $\lg w_D(\omega)$ is so flattened out that the inclusion of 2ω, even at $\omega = 10\max\{k_1; k_2\}$, does not result in a noticeable change in the function. Let us demonstrate this. Usually, the lowest proper frequencies of shock stands are on the order of 10^4 rad/s, and the maximum promising spectrum frequencies in practical terms constitutes 10^5 rad/s. If $2\omega = 9$ $(k_1 + k_2)$, then $\lg(2\omega + k_1 + k_2) = \lg(10(k_1 + k_2)) = \lg 10 + \lg(10^4) = 1 + 4$. If ω is smaller, then the first summand will be less than 1. Log dimension values should not concern us, since the log sum collapsing into log multiplications yields non-dimensional values. Let us write the expression for the shock spectrum in dimensionless values by dividing (16) by acceleration of the free fall g:

$$S(\omega) = \lg\left(\frac{w_D}{g}\right). \tag{19}$$

Let us put expression (17) into (19) in light of the above (18):

$$S(\omega) = \begin{cases} \lg\left(\frac{vp_3^2}{g|k_1-k_2|}\right) + 2\lg\omega - 2\lg\sqrt{k_1 k_2}, & \omega \le \sqrt{k_1 k_2}, \\ \lg\left(\frac{vp_3^2}{g|k_1-k_2|}\right), & \omega \ge \sqrt{k_1 k_2}. \end{cases} \tag{20}$$

Conjugating frequency

$$\omega_0 = \sqrt{k_1 k_2} = \sqrt{4p_1^2 p_2^2} \tag{21}$$

is determined from the condition of intersection of lines (20). In Fig. 2, the dotted broken line shows the shock spectrum obtained from the introduced calculation (20) with initial data coinciding with the aforementioned experiment (Fig. 1).

Continuum Model

To describe elastic deformations of the table and the impactor, we will use the stand model designed as a straight homogeneous rod with longitudinal oscillations. Figure 6 shows the adopted dynamic model of the shock process in the stand, where h is the table length, l-h is the projectile length and v is the vector of initial projectile velocity. The equation of longitudinal oscillations of the rod is as follows:

$$\frac{\partial^2 u}{\partial t^2} - a^2 \frac{\partial^2 u}{\partial x^2} = 0, \quad a = \sqrt{\frac{E}{\rho}}, \tag{22}$$

with boundary

$$u\big|_{x=0} = 0, \quad \frac{\partial u}{\partial x}\bigg|_{x=l} = 0 \tag{23}$$

and initial

$$u\big|_{t=0} = 0, \quad \frac{\partial u}{\partial t}\bigg|_{t=0} = \begin{cases} 0, & x \in [0; h] \\ -v, & x \in [h; l] \end{cases} \tag{24}$$

conditions, where u is the displacement of the rod with initial coordinate x caused by rod deformation, a is the speed of wave distribution in the rod, and E and ρ are the normal elasticity modulus and the density of the rod material. Let us apply the Fourier method. We will try to solve Eq. (22) in the following form:

$$u(x,t) = \sum_{n=0}^{\infty} X_n(x) T_n(t). \tag{25}$$

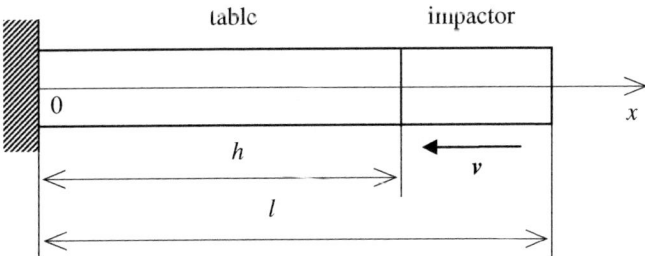

Fig. 6 Continuum model of the shock stand

The function that corresponds to the boundary conditions (23) is

$$X_n(x) = \sin\left(\frac{(2n+1)\pi x}{2l}\right), \tag{26}$$

and the function that corresponds to the first initial condition (24) is

$$T_n(t) = b_n \sin\left(\frac{(2n+1)\pi a t}{2l}\right). \tag{27}$$

Let us put (26) and (27) into (25) and then into the second initial condition (24). If we multiply both parts of the obtained congruence by $X_n(x)$ and integrate them into the interval from 0 to l, due to the orthogonality of our own forms of (26), we will obtain

$$
\begin{aligned}
b_n &= \frac{4v}{(2n+1)\pi a} \int_h^l \sin\left(\frac{(2n+1)\pi x}{2l}\right) dx \\
&= \frac{8vl}{(2n+1)^2 \pi^2 a} \cos\left(\frac{(2n+1)\pi h}{2l}\right).
\end{aligned}
\tag{28}
$$

Let us put (28), (27) and (26) into (25):

$$
\begin{aligned}
u(x,t) = \sum_{n=0}^{\infty} \frac{8vl}{(2n+1)^2 \pi^2 a} \cos\left(\frac{(2n+1)\pi h}{2l}\right) \\
\sin\left(\frac{(2n+1)\pi a t}{2l}\right) \sin\left(\frac{(2n+1)\pi x}{2l}\right).
\end{aligned}
\tag{29}
$$

The function of table movement is defined as the displacement of its middle:

$$
\begin{aligned}
x_{\tilde{C}} = \sum_{n=0}^{\infty} \frac{8vl}{(2n+1)^2 \pi^2 a} \cos\left(\frac{(2n+1)\pi h}{2l}\right) \\
\sin\left(\frac{(2n+1)\pi a t}{2l}\right) \sin\left(\frac{(2n+1)\pi h}{4l}\right).
\end{aligned}
\tag{30}
$$

Due to the quick convergence of range (30), let us confine ourselves to the first summand at $n = 0$. Then,

$$x_{\tilde{C}} = a_m \sin(k_3 t), \text{ where } a_m = \frac{8vl}{\pi^2 a} \cos\left(\frac{\pi h}{2l}\right) \sin\left(\frac{\pi h}{4l}\right); \ k_3 = \frac{\pi a}{2l}. \tag{31}$$

Let us differentiate the last equality twice and show it in Fig. 7.

Fig. 7 Dependence between shock acceleration (g) and time (t) in homogeneous rod stand modeling

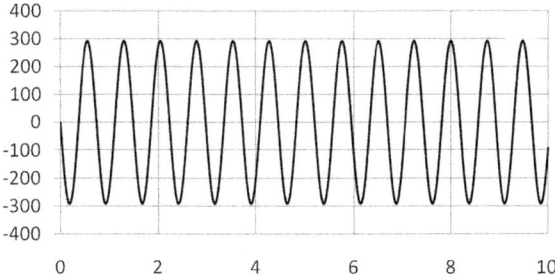

While this graph is dramatically different from the graphs in Figs. 1 and 5, its shock spectrum, calculated using the method given below, is almost indistinguishable from the one obtained using the accurate method (the solid curve in Fig. 2) and the discrete model (20) (the dotted broken line in Fig. 2). Let us demonstrate this point. Let us take the Laplace transform of table movement (31)

$$X_{\hat{C}} = a_m \frac{k_3}{p^2 + k_3^2}, \tag{32}$$

put it into (2), and decompose it into partial fractions:

$$X_D = \frac{\omega^2}{p^2 + \omega^2} a_m \frac{k_3}{p^2 + k_3^2} = \frac{a_m k_3 \omega^2}{k_3^2 - \omega^2} \left(\frac{1}{p^2 + \omega^2} - \frac{1}{p^2 + k_3^2} \right). \tag{33}$$

Let us record the inverse Laplace transform of (33):

$$x_D = \frac{a_m k_3 \omega^2}{k_3^2 - \omega^2} \left[\frac{1}{\omega} \sin(\omega t) - \frac{1}{k_3} \sin(k_3 t) \right]. \tag{34}$$

After double differentiation in time (34) on the basis of assertions that precede (16), let us write the equality

$$w_D(\omega) = \frac{a_m k_3 \omega^2}{|k_3^2 - \omega^2|} (\omega + k_3) = \frac{a_m k_3 \omega^2}{|k_3 - \omega|}. \tag{35}$$

This expression was rewritten in a form analogous to (17) and (18). For low ω frequencies, we can take

$$w_{D1}(\omega) = a_m \omega^2, \tag{36}$$

and for high,

$$w_{D2}(\omega) = a_m 4k_3^2. \tag{37}$$

The last equality corresponds to the local minimum of function (35) at $\omega = 2k_3$. Approximation (36) passes through the same point. Let us write the expression for the shock spectrum obtained on the basis of the continuum stand model:

$$S(\omega) = \begin{cases} \lg\left(\frac{a_m}{g}\right) + 2\lg\omega, & \omega \le 2k_3 \\ \lg\left(\frac{a_m}{g}\right) + 2\lg 2k_3, & \omega \ge 2k_3 \end{cases} \tag{38}$$

In Fig. 2, the angled solid line with markers shows the shock spectrum obtained on the basis of the solution of the continuum model of the test stand and its rough estimate (38).

Conclusion

The article develops and explains approximate estimates of shock spectra accelerations reproduced on test stands. The obtained formulas facilitate a directed search of standard shock stand sizes, choice of gaskets, impact velocity and other dynamic parameters used to create shock pulses with the desired spectrum.

References

1. Chang KY (2002) Pyrotechnic devices, shock levels and their applications. In: 9th International congress on sound and vibration Orlando, USA, July 2002, 19 pp
2. Lee J-R, Chia CC, Kong C-W (2012) Review of pyroshock wave measurement and simulation for space systems. J Measure 45:631–642
3. Andrienko PA, Karazin VI, Hlebosolov IO (2012) About combined effects tests. In: Modern mechanical engineering: science and education, no 2, pp 142–149
4. Evgrafov AN, Karazin VI, Smirnov GA (1999) Rotary stands for the reproduction of motion parameters. Scientific and technical bulletins of St-Petersburg Polytechnic University, no 3, pp 89–94 (rus)
5. Evgrafov AN, Karazin VI, Hlebosolov IO (2003) Reproduction of the motion parameters on rotary stands. In: The theory of mechanisms and machines, vol 1, no 1, pp 92–96 (rus)
6. Karazin VI, Kolesnikov SV, Litvinov SD, Sukhanov AA, Hlebosolov IO (2013) Peculiarities of the simulation and reproduction of vibration impact. The theory of mechanisms and machines, vol 11, no 22, pp 55–64 (rus)
7. Karazin VI, Kolesnikov SV, Litvinov SD, Sukhanov AA, Hlebosolov IO (2013) The parameters optimization of broadband vibro-impact mechanical stand. In: Modern mechanical engineering: science and education, 752 pp (rus)
8. Yarovitsyn VS, Litvinov SD, Karazin VI, Sukhanov AA, Hlebosolov IO (2007) A device for testing products for vibro-impact loads. Invention patent (RUS 2348021)

9. Komarov IS. Ground-based experimental testing of rocket-space equipment products to impact from pyrotechnic separation means. TsNIIMash. Electronic journal "Works of Moscow Aviation Institute". Issue № 71, 22 p. www.mai.ru/science/trudy/ (rus)
10. GOST R 51371-99 (2000) Test methods for resistance to mechanical external influencing factors of machines, instruments and other technical products. Impact tests–Moscow, 24 pp (rus)
11. Timofeev EG, Zhukov IA (2016) To the development of a numerical method for studying shock processes in the rod system of impact machines. In: Modern mechanical engineering: science and education MMESE-2016, St-Petersburg, Publishing House of the Polytechnic University, pp 540–549 (rus)
12. Pervozvansky AA (2010) The theory of automatic control. Tutorial. 2nd edn. St-Petersburg, Publishing House "Lan", 624 pp (rus)
13. Babakov IM (1958) Oscillation theory. State Publishing House of Technical and Theoretical Literature–Moscow, 628 pp (rus)
14. Timoshenko SP., Young D.H., Weaver U (1985) Fluctuations in engineering. Translated from English into Russian–Moscow, Mechanical Engineering, 472 pp

Some Peculiarities of Electric Drive Impact on the Dynamics of Cyclic Machines

Iosif I. Vulfson

Abstract The article is devoted to the study of vibrations in the drives of machines with cyclic mechanisms, taking into account the characteristics of the electric motor. It is established that with the simplified modeling often used in engineering practice, when the effect of high equivalent compliance of the motor is ignored, significant errors in the determination of resonant frequencies are possible. A number of new modifications of models for one-sided and two-sided drives are proposed, as well as for preliminary engineering estimates. An engineering technique for calculating similar systems is described, based on the use of transition matrices well suited to computer procedures. The results of experiments and computer simulations are presented.

Keywords Electric drive · Cyclic machines · Vibrations

Introduction

Modern technological machines are complex systems in which three functional parts can be distinguished: the engine, the mechanical system (the working machine) and, in some cases, the motion control system. In cyclic technological machines, which include many from the textile, light, printing and other industries, the movement of the actuators is usually realized in the mechanical system itself, without the use of special software controls. This is due to the higher requirements for accuracy of positioning of the working elements in performing complex technological operations.

Dynamic processes arising from the influence of the electric motor are manifested not only in the change in the force and amplitude-frequency characteristics, but also in the occurrence of significant dynamic errors in the reproduction of a

I. I. Vulfson (✉)
Saint Petersburg State University of Industrial Technologies and Design,
Bolshaya Morskaya Str.18, 191186 Saint Petersburg, Russia
e-mail: jvulf@yandex.ru

© Springer International Publishing AG 2018
A. N. Evgrafov (ed.), *Advances in Mechanical Engineering*, Lecture Notes
in Mechanical Engineering, https://doi.org/10.1007/978-3-319-72929-9_17

given programmed motion. When determining the level of vibroactivity for steady-state regimes on the "input" of the drive of machines, a constant average angular velocity is usually applied. In this case, the forced movement of the drive with this speed is usually treated as an assumption, according to which the oscillatory drive system at the input can be considered as fixed. It is of interest as to the extent to which this assumption is correct.

The description of electromagnetic oscillatory processes in engines and their influence on the dynamics of machines is associated with the solution of rather complex systems of nonlinear differential equations [1–3]. Therefore, when solving engineering problems, the criteria for assessment play an important role, allowing us to use decomposition and aggregation [4, 5] to construct effective dynamic models in a given class of systems. Among the monographs devoted to the analysis of the machine set that take into account the dynamic characteristics of the electric motor, we shall single out the works of MZ Kolovsky and VL Weits, which are closest to the solution of practical engineering problems [2,6,7]. With respect to steady-state regimes in engineering practice, approximate linearized equations have proved to be well-suited for the DC motor and asynchronous motors, and have the form

$$\Omega_m = \Omega_{m0}[1 - v_m(M_m + \tau_m\dot{M})], \tag{1}$$

where M_m is the motor torque; v_m is the steepness coefficient of static characteristics; τ_m is the electromagnetic time constant; and Ω_{m0} is the angular velocity of ideal idling $(M_m \equiv 0)$.

For asynchronous motors $\tau_m = (2\pi f_c s_\kappa)^{-1}$; $s_\kappa = (1 - \Omega_m^n/\Omega_{m0})(\xi + \sqrt{\xi^2 - 1})$; $v_m = s_\kappa/(2M_m^n\xi)$. Here, s_κ is the critical slip; $f_c = 50$ Hz is the frequency of the network; and ξ is the ratio between the maximum torque and the rated torque.

The characteristic of the electric motor (1) corresponds to the Maxwell rheological model, in which the rotor is connected to the stator by means of an "elastic element" with a stiffness coefficient $c_0 = (v_m\Omega_{m0}\tau_m)^{-1}$ and a series-connected damper with a resistance moment $b_0\Omega_m$ at $b_0 = (v_m\Omega_{m0})^{-1}$ [8, 9]. Applied to the drives of cyclic machines, the research on this problem is reflected in a number of publications [10, 11, 6, 7, 12, 13], and is further developed in this work. When dissipative forces are taken into account, a laconic engineering method for investigating this problem based on the method of complex amplitudes was proposed by Sorokin [14].

Dynamic Analysis of Machines with One-Way Drive

Figure 1 shows a number of typical dynamic models that serve as a basis for carrying out not only analysis but also dynamic synthesis of oscillatory systems in a given class of problems. Here, the following symbols are used: J_0, J_1, J_2 are

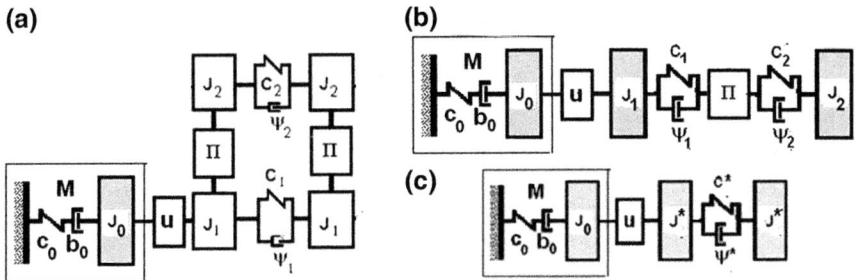

Fig. 1 Typical dynamic models: **a** Ring structure model (1); **b** chain structure model (2); **c** generalized model (3)

moments of inertia; Π is an operator corresponding to a nonlinear position function (see below); c, ψ are the reduced drive stiffness and dissipative factors; and u is the gearing ratio. On models, sub-systems **M** and **u** corresponding to the engine and gearbox are allocated.

The model of the ring structure (Fig. 1a) is very common in machines with an increased length of their process zone (for example, in a number of textile and printing machines), and in machines with massive actuators (for example, in flat printing machines, machine-tools, etc.). It can be shown that with a certain choice of parameters c_2^* and J_1^*, J_2^* (see below), both models are represented by the model shown in Fig. 1c.

On the "input" and "output" of cyclic mechanisms, the relationship between the coordinates is described by a nonlinear position function. For an ideal mechanism, one for which there are no gaps and all links are taken as absolutely rigid, $\varphi = \varphi_* = \bar{\omega}t$, where $\bar{\omega}$ is the average angular velocity of the input link. Let, $\varphi = \varphi_* + \Delta\varphi$, where $\Delta\varphi$ is the dynamic error arising at oscillations. As the ideal position function is continuous and differentiable, we can linearize this function in the vicinity of the program motion: $\Pi(\varphi_* + q) \approx \Pi(\varphi_*) + \Pi'(\varphi_*)q$, where $\Pi' = d\Pi/d\varphi$ is the first geometrical transfer function of the mechanism (the analog of velocity). Thus, for small oscillations that are incommensurable with the "ideal" coordinates, we have replaced the nonlinear constraint with one that is nonstationary with practically no damage to accuracy. Nonstationary constraints can be an important source of increasing vibroactivity in cyclic mechanisms. In particular, in this case, the effective dissipation coefficient corresponding to nonlinear positional dissipative forces decreases significantly and may even become negative, which leads to an increase in the vibration amplitudes, as well as to the possibility of instability of dynamical regimes under parametric excitation [15,8, 9, 10, 11,6,7,12, 13].

We analyze the structure of the first geometric transfer function in more detail. Let $\Pi' = \Pi_0' + \Pi_v'$, where the terms correspond to slow and fast motions. For certainty, we take $\Pi'(\varphi) = h[\sin\varphi + \varepsilon\sin(v\varphi + \gamma)]$ for $h = 1$. Thus, the second term here describes the "fast" harmonic with number of frequency v and relative depth of pulsation ε.

We consider two special cases. First, we temporarily exclude from consideration the influence of the electric motor and take the angular velocity of the main shaft to be equal: $\omega = \bar{\omega} = $ const. We note that for the oscillatory system, this assumption corresponds to a rigid fixation at the "input". Let us give dynamical models for the form of the chain structure model (Fig. 1c). It can be shown that for model 1 (see Fig. 1a), $c^* = \tilde{c}_1(1+\zeta\Pi'^2)$, $J^* = J_1(1+\mu\Pi'^2)$, where $\zeta = c_2/c_1$; $\mu = J_2/J_1$; $\tilde{c}_1(\delta_1^*) = c_1(1+2i\delta_1^*)$; $\delta_1^* = \psi_1^*/(4\pi)$; $\psi^* = (\psi_1 c_1 + \psi_2 c_2)/(c_1+c_2)$; for model 2 (see Fig. 1b), $\psi^* = \psi_2^* = |c^*|(\psi_1/c_1 + \psi_2/c_2)$; $\delta_2^* = \psi_2^*/(4\pi)$. Here, $c^* = \tilde{c}_1(\delta_2^*)/(\Pi'^2 + \zeta^{-1})$, $\tilde{c}_1(\delta_2^*)$ is the complex form of an elastic-dissipative element.

In the framework of the problem posed, when illustrating the results of the analysis, we use model 1, which, after the above linearization in the vicinity of the program motion, is described by differential equations (for details, see below):

$$a(\varphi)\ddot{q} + 2\mu J_1 \Pi'_* \Pi''_* \omega \dot{q} + \tilde{c}_1(\delta_1^*)(1+\zeta\Pi'^2_*)q = M(t)\Pi'_* - \mu J_1 \Pi'_* \Pi''_* \omega^2, \qquad (2)$$

where $a(\varphi) = J_1(1+\mu\Pi'^2_*)$; $M(\varphi_*)$ is the external torque applied to the output link; and q is dynamic error of the working element.

This equation corresponds to the following homogeneous differential equation with variable coefficients:

$$\ddot{q} + 2n(t)\dot{q} + p^2(t)q = 0. \qquad (3)$$

Here, $p(t) = k\sqrt{(1+\zeta\Pi'^2_*)/(1+\mu)\Pi'^2_*}$; $n(t) = n_0(t) + n_h(t)$, where $k = \sqrt{c_1/J_1}$; $n_0(t) = \delta p(t)$; $\delta \approx \psi/(4\pi)$; $n_h(t) = \omega(da/d\varphi)/(2a)$.

The functions n_0 and n_h describe the dissipative and gyroscopic components, and the function $p(t)$ is the variable "natural" frequency. Figure 2 shows the graphs $p_0(t)$ (bold line) and $p(t)$ (thin line) corresponding to slow and fast changes in functions $\Pi'(\varphi_*)$ (see above).

In [8], on the basis of the method of the conditional oscillator, it was shown that the suppression of parametric excitation is achieved when

$$\vartheta = 2\pi\delta > \vartheta^* = \pi p_0^{-1} |a'/a + p'/p - p_0'/p_0|, \qquad (4)$$

where ϑ is the logarithmic decrement; and $p_0(t)$ corresponds to the slowly varying component of the function $p(t)$.

Since $a(0) = a(2\pi)$, this function does not affect the threshold conditions for the excitation of parametric resonance. However, within the cycle, zones are possible in which the amplitude of the oscillations increases, often leading to significant dynamic errors in the realization of the programmed motion. With a slow change in the parameters, the condition for an arbitrary time interval is satisfied if

Fig. 2 "Natural" frequencies without taking the engine into account: (1) $\mu = \zeta = 0.2$; (2) $\mu = 0.2, \zeta = 0$ (3) $\mu = 0, \zeta = 0.2$

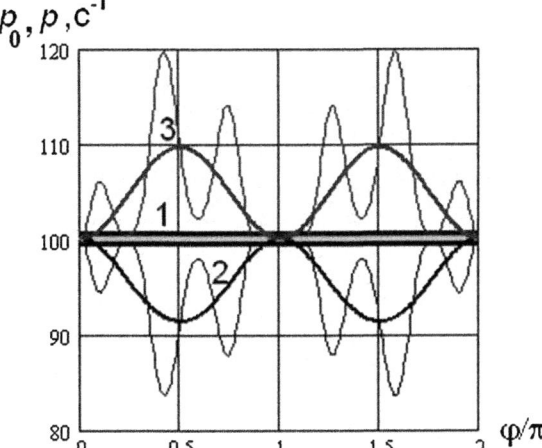

$$\delta_0 > \delta^*(\varphi) = \delta_h(\varphi) + \delta_p^*(\varphi), \tag{5}$$

where $\delta_p^*(\varphi) = 0.5 p_0(\varphi)^{-2} \frac{dp_0(\varphi)}{d\varphi}$.

The function δ_h corresponds to gyroscopic forces and $\delta_p^*(\varphi) = 0.5 p_0(\varphi)^{-2} \frac{dp_0(\varphi)}{d\varphi}$ reflects the influence of the variability of the "natural" oscillation frequency. The maximum value of the function corresponds to the critical excitation level. Inequality (5) coincides with the results obtained on the basis of the direct Lyapunov method, which establishes a sufficient condition for dynamic stability [7, 3, 5]. It can be shown that with slowly varying parameters, it is also a necessary condition. As applied to this problem, condition (5) can be qualified as a quasi-stationary condition [9]. In spite of the variability of the system parameters, the "natural" frequencies in the first approximation retain a constant value. In this case, there is no growth of amplitudes in the interval of the kinematic cycle. The realization of the quasi-stationary conditions is a very effective way of eliminating the possibility of excitation of parametric resonances and a decrease in the vibroactivity of the system.

In the second special case, we assume that $c_1 \to \infty$, i.e., the elastic properties of the drive are not now taken into account. Then, again, when representing the angular velocity of the motor as a sum of constant and variable components, after linearization in the vicinity of the programmed motion, we obtain

$$\Delta\ddot{\omega}_m + 2n(\varphi_*)\Delta\dot{\omega}_m + k^2(\varphi_*)\Delta\omega_m = W(\varphi_*), \tag{6}$$

where $\Delta\omega_m$ is the dynamic speed error; $2n = \tau_m^{-1} + 2J'\bar{\omega}/J$; $J = J(\varphi)$; $k^2 = J^{-1}[(v_m\tau_m\bar{\omega}_m)^{-1} + \bar{\omega}J'\tau_m^{-1}]$; and $W = -J^{-1}(\Delta M\tau_m^{-1} + \Delta M'\bar{\omega} + 0.5\bar{\omega}^2 J'\tau_m^{-1})$.

The dash on top corresponds to the average value on the period. For model 1, the variable reduced moment of inertia is $J(\varphi) = 2J_1(1 + \mu\Pi_*^{'2})$ and $\Delta M = M - \bar{M}$.

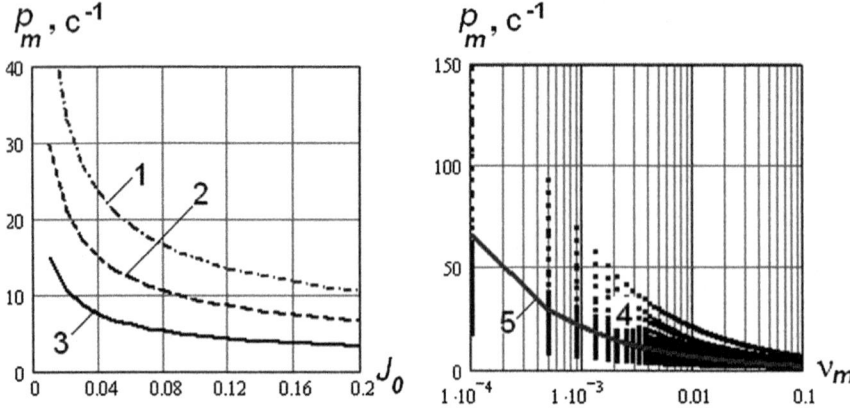

Fig. 3 Graphs $p_m(J_0), p_m(v_m)$:   (1) $v_m = 0.002$;   (2) $v_m = 0.005$;   (3) $v_m = 0.02$;
(4) $J_0 = 0.001\ldots0.02$; (5) $J_0 = 0.1$

In Fig. 3, the graphs of the dependence of the partial frequency on the inertial characteristics of the drive and the coefficient of steepness of the static characteristics of the engine are given for model 1.

To determine the amplitude-frequency characteristics, taking into account the characteristics of the engine and the mechanical drive together, we use a laconic method based on transition matrices. For the considered chain system (see Fig. 1c), the transition matrix has the form.

$$\Gamma = \Gamma_{J*}\Gamma_{c*}\Gamma_{J*}\Gamma_u\Gamma_{J0}\Gamma_M \tag{7}$$

Here,

$$\Gamma_{Js} = \begin{pmatrix} 1 & 0 \\ -J_s\omega^2 & 1 \end{pmatrix}; \quad \Gamma_{c*} = \begin{pmatrix} 1 & c*^{-1} \\ 0 & 1 \end{pmatrix}; \quad \Gamma_u = \begin{pmatrix} u & 0 \\ 0 & u^{-1} \end{pmatrix};$$

$$\Gamma_M = \begin{pmatrix} 1 & c_0^{-1} - i(b_0\Omega_{\text{Д}})^{-1} \\ 0 & 1 \end{pmatrix}; \quad i = \sqrt{-1}.$$

As follows from the right-hand side of Eq. (2), the forcing moment arising on the working organ is composed of a kinematic perturbation, depending on its acceleration under ideal motion, and a power disturbance from technological forces and resistance forces. We emphasize that the peculiarity of the cyclic mechanisms is the transformation of the constant components at the "input" into the variable functions at the "output".

Let the driving moment applied to the output link be $M = \sum_r M_r \sin(\omega_r t + \alpha_r)$. (To simplify the record, the subscript below is omitted everywhere else.) On the basis of (7),

$$\begin{pmatrix} \tilde{A} \\ \tilde{M}_{1r} \end{pmatrix} = \begin{pmatrix} \tilde{g}_{11} & \tilde{g}_{12} \\ \tilde{g}_{21} & \tilde{g}_{22} \end{pmatrix} \begin{pmatrix} 0 \\ \tilde{M}_{mr} \end{pmatrix}, \tag{8}$$

where $\tilde{g}_{kj}$ are the elements of the matrix $\mathbf{\Gamma}$; $\tilde{A}$ is the amplitude of the output link; and $\tilde{M}_{1r}, \tilde{M}_{mr}$ are the amplitudes of the motor torque (as above, the wave indicates the complex shape of the function).

From (7) and (8), it follows that

$$\tilde{M}_m = \beta_{22}(\omega, \varphi)\tilde{M}_1; \quad \tilde{A} = \beta_0 \tilde{M}_1. \tag{9}$$

Here, $\beta_{12}(\omega, \varphi) = |\tilde{g}_{12}|; \beta_{22}(\omega, \varphi) = |\tilde{g}_{22}|^{-1}; \beta_0(\omega, \varphi) = \beta_{12}(\omega, \varphi)/\beta_{22}(\omega, \varphi)$ (Fig. 4).

For the model under consideration, the resonance regimes correspond to the minima of the following functions: $\beta_{22}(\omega, \varphi)$—for a fixed input and a free output; $\beta_{12}(\omega, \varphi)$—at both fixed ends of the oscillating circuit. It should be borne in mind that, because of the slow change in the position function (region 0), the minima of these functions are "floating" in nature. This can lead to the appearance of a beating mode, and in some cases, to local violations of the conditions of dynamic stability as well.

Figure 5 shows the typical frequency-amplitude (AFC) and phase-frequency (PFC) characteristics obtained on the basis of (8) and (9): $A = |\tilde{A}| \approx \operatorname{Re} \tilde{A}$; $\gamma = \operatorname{Im} \tilde{A}$; $\Omega_m^n = \omega/u$; $v_m = 0.003(\mathrm{H_M})^{-1}$; $\delta = 0.03$. We draw attention to the fact that the AFC is proportional to the function β_0 only for a constant amplitude of the angular momentum M_1. With kinematic excitation, the amplitude is proportional $\beta_0 \omega^2$ (Fig. 5).

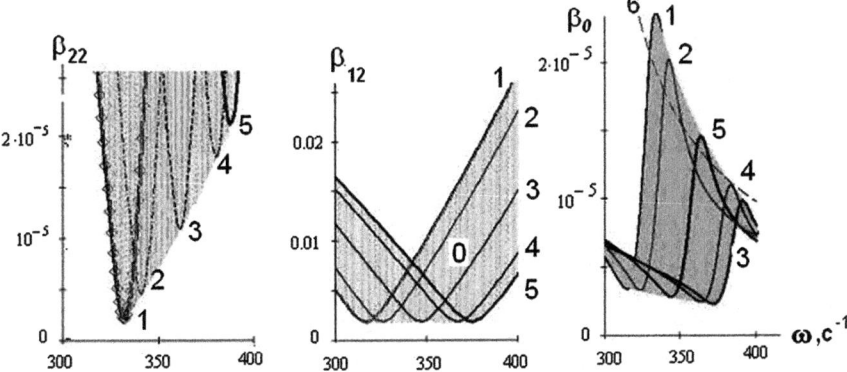

Fig. 4 Graphs of functions $\beta_0, \beta_{12}, \beta_{22}$: (1) $\varphi = 0$; (2) $\varphi = \pi/8$; (3) $\varphi = \pi/4$; (4) $\varphi = 3\pi/4$; (5) $\varphi = \pi/2$; (0) $0 \leq \varphi \leq \pi/2$; about curve (6) see below

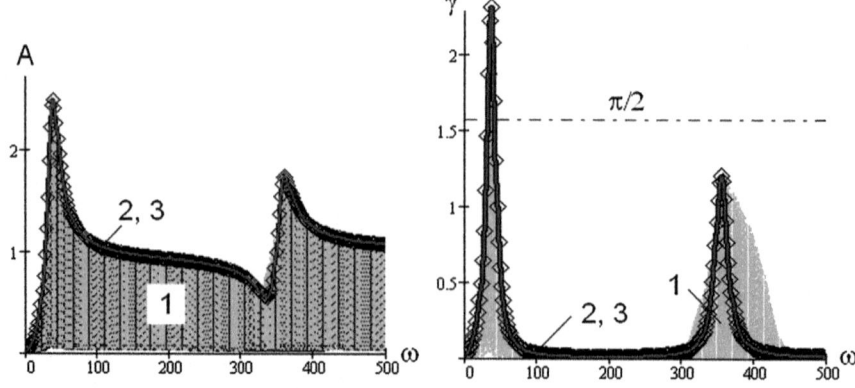

Fig. 5 Graphs of AFC and PFC: (1) $0 \leq \varphi \leq \pi$; (2) $\varphi = 0, \pi \, (\zeta = \mu)$; (3) $\varphi = 0, \pi \, (\zeta \neq \mu)$

The averaged partial frequencies for the initial data received are equal: $k_1 = 39.6$; $k_2 = 100 \, s^{-1}$. When connecting the motor to the drive, the averaged low natural frequency differs little from the partial frequency of the engine. At first glance, the significant excess of the second resonant frequency $360 \, s^{-1}$ in comparison with the natural frequency, without taking the motor into account $p0 = 100 \, s^{-1}$ (see Fig. 2), causes some bewilderment and seems implausible. This is due to the fact that the revealed effect does not agree with the traditional opinion when the role of the engine is limited, in practical terms, only by the appearance of an additional natural frequency, on which the so-called electromagnetic resonance arises.

Let us dwell on this issue in more detail, since it is associated with significant errors in the dynamic analysis and synthesis of machine drives. Usually, for the considered class of engines, $\Delta\Omega = \Omega_{m0} - \Omega_m^n \approx (5-10)s^{-1}$; $\tau_m \approx (0.01-0.15)s$. As shown in [6], under these conditions, the equivalent dynamic error at the input of the drive under the action of the nominal torque corresponds to deformation of the drive $(0.05-1.4)$rad, which, as a rule, considerably exceeds the deformations of the shafting line. This justifies the analysis of the limiting case, when the equivalent "rigidity" of the engine is zero. In other words, the engine is "disconnected" from the power network, and its role is limited to taking into account the moment of inertia of the rotor. In this case, the dynamic model (see Fig. 1b) displays a system with two degrees of freedom with one cyclic coordinate and a sole natural frequency, different from zero (Fig. 6):

$$p_*(\varphi) = \sqrt{[2J^*(\varphi_*) + J_{01}]c(\varphi)/\{[J^*(\varphi) + J_{01}]J^*(\varphi)\}}, \tag{10}$$

where $J(\varphi) = J_1(1 + \mu\Pi_*^{\prime 2})$; $J_{01} = J_0 u^{-2}$.

The resonant regimes in the linear approximation correspond to the "natural" frequencies, which are the roots of the biquadratic equation

Fig. 6 Graphs $p_*(\varphi)$:
(1) $\xi = 0$; (2) $\xi \neq 0$; (3) the
boundaries of the frequency
range when changing $p_2(\varphi)$
(see Fig. 3, line 6); $u = 1$

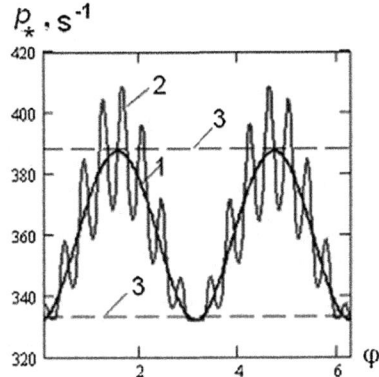

$$f_2(\varphi_*)p^4 - f_1(\varphi_*)p^2 + f_0(\varphi_*) = 0. \tag{11}$$

Here,

$$f_0(\varphi_*) = c(\varphi_*)/(v_m\tau);$$
$$f_1(\varphi_*) = [(v_m\tau)^{-1} + c(\varphi_*)]J(\varphi_*) + (J_{01} + J(\varphi_*))c(\varphi_*);$$
$$f_2(\varphi_*) = [J_{01} + J(\varphi_*)]J(\varphi_*).$$

The second resonant frequency is usually significantly higher than the partial frequency of the mechanical subsystem of the drive (see Figs. 2 and 3). This frequency, besides other parameters, largely depends on the ratio of the moments of inertia $\chi = J_1/J_{01} = J_1u^2/J_0$ (Fig. 7).

It should be noted that this effect indicates the need to correct the commonly accepted methods for assessing the role of an electric motor on the basis of simplified dynamic models. In particular, the simplified model shown in Fig. 1b, in which the selected subsystem M corresponding to the electric motor is replaced by a restraint of the vibration chain, is in frequent use. This introduces a significant error

Fig. 7 Graphs $p_2(\chi)$ at
$\varphi = \pi/2$: (1) $\mu = \zeta = 0.4$;
(2) $\mu = 0$, $\zeta = 0.4$;
(3) $\mu = 0.4$, $\zeta = 0$

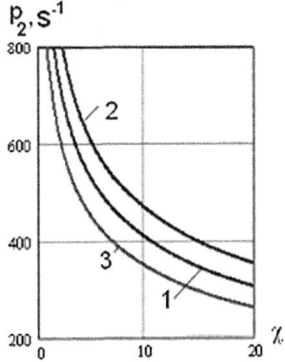

in determining the resonance frequency. For example, in the considered system, the resonance frequency p_2 is approximately 3.5 times higher than the frequency p_0 corresponding to this simplification. The graph $p_2(\chi)$ in Fig. 7 indicates that this error can still be much higher. At the same time, taking into account another simplification $c_0 = 0$ (free end) and retaining the element corresponding to the motor rotor in the model, we have a discrepancy between the frequencies and less than 10% (see Fig. 4, curve 6). Thus, with the frequently used simplified modeling, taking into account the electric motor, it is necessary to exclude the closed input end, while retaining only the inertial characteristics of the motor rotor.

A feature of the AFC and PFC is the variability of amplitudes and phase shifts within the kinematic cycle, which is caused by a slow change of the component of the "natural" frequency $p_0(t)$ (see Fig. 4, region 0). This effect leads to the appearance of beat modes, which was confirmed by both computer simulation and experimental studies. If the parameters are changed slowly, a local transient mode is implemented. On the graphs of the frequency characteristics (Fig. 8), constructed on the basis of (10) and (11), for $J_0 = 0.01\,\text{kg m}^2$, $J_1 = 0.4\,\text{kg m}^2$, $u = 1$; solid lines correspond to Π'_0 ($\varepsilon = 0$), and dashed lines to Π'_v ($\varepsilon = 0.5$).

The analysis of the graphs shows that the electromagnetic frequency p_1 responds sensitively to the inertial characteristics (parameter μ) and weakly to the elasticity of drive (parameter ζ). The frequency p_2, which plays the dominant role in the formation of forced and parametric vibrations of the actuators, essentially depends on the ratio of the parameters ζ and μ. In our case, the quasi-stationary conditions correspond to $\zeta = \mu$ (curves 1, dashed lines). In this case, the pulsation of the parameters, which is caused by non-stationary constraints, and, consequently, the possibility of excitation of parametric resonances, is practically eliminated. The graphs show a significant decrease in the pulsation of this frequency at $\zeta = \mu$. This confirms the effectiveness of quasistationarity in the optimization of dynamic characteristics [9]. The graph p_* essentially duplicates the graph for frequency p_2, which indicates a large equivalent "compliance" of the engine and a relatively small effect of its electromagnetic characteristics on this frequency.

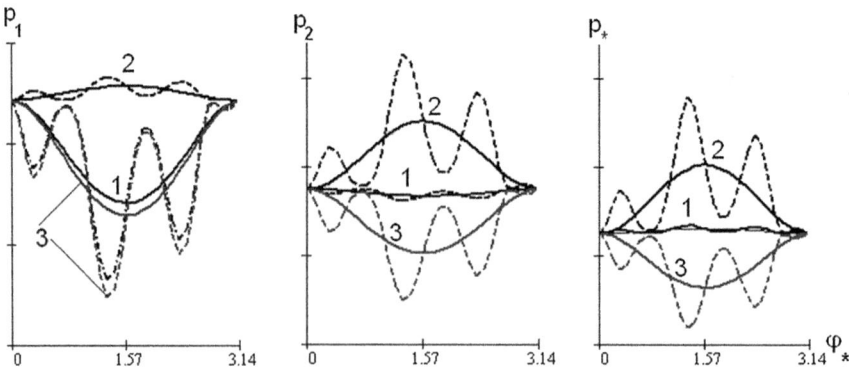

Fig. 8 Graphs $p_1(\varphi_*), p_2(\varphi_*), p_*(\varphi_*)$ (1) $\mu = \zeta = 0.4$; (2) $\mu = 0, \zeta = 0.4$; (3) $\mu = 0.4, \zeta = 0$

Above, with kinematic excitation for determining the moment of resistance, we proceeded from the assumption that the moment of inertial forces of the output link is determined by a given law of program motion at a constant angular velocity of the input link. This moment at $\omega = $ const corresponds to the second term on the right-hand side of Eq. (2): $M^* = -\mu J_1 \Pi'_* \Pi''_* \omega^2$. However, firstly, when $\omega \neq$ const, we need to adjust the acceleration of the driven link:

$$M^* = -\mu J_1 \Pi' (\Pi'' \omega^2 + \Pi' d\omega/dt) \qquad (12)$$

Secondly, the function $\omega(t)$, along with the other factors, also now depends on M^*. To identify the qualitative features of the effects that arise in this case, we use computer modeling based on a system of differential equations that combines the mechanical and electromechanical characteristics of the system:

$$\left.\begin{array}{l} \Delta\omega' + (\Delta M + \Delta M' \tau_m \Omega_m^n u)\Omega_{m0}(1 - s_0)v_m = 0 \\ [1 + 0.5\bar{p}_0(1 - \cos 2\varphi)]\Delta\omega' \Omega_m^n u^{-1} \\ + 0.5\bar{p}_0 u^{-1} \sin 2\varphi(\Omega_m^n + \Delta\omega)^2 - \Delta\omega J_0^{-1} = r\cos\varphi; \end{array}\right\} \qquad (13)$$

$$\Omega_m^n \Delta\varphi' - \Delta\omega = 0,$$

where $\bar{p}_0$ is the rms value of the function $p_0(\varphi)$.

Figure 9a shows the graphs of relative angular velocity deviations (curve 1) obtained on the basis of (13) and the corresponding phase shifts of the angles of rotation of the input link (curve 2).

The graphs clearly show the decrease in speed in the working and reverse motion, the increase in speed in the braking areas, as well as the phase shifts corresponding to these sections. In this case, one has to deal with the effect that occurs when the argument of the geometric characteristics of the cyclic mechanism is significantly different from the linear function, which can lead to significant distortions in the program motion and a change in the torque on the motor. Figure 9b shows the transformation of the moment that arises from these distortions, taking into account the static and dynamic characteristics of the engine. The

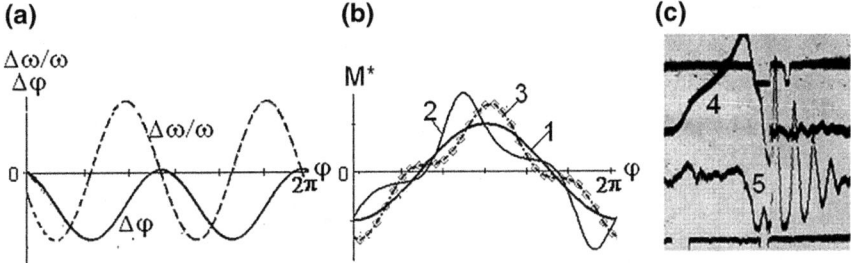

Fig. 9 Distortions of kinematic and dynamic characteristics: (1) $\omega = $ const; (2) $\tau_m = 0$; (3) $\tau_m \neq 0$; (4) speed; (5) acceleration

mitigating effect of the motor dynamic characteristic in comparison with the static one is of interest. The latter is related to the influence of the electromagnetic time constant, which suppresses drastic changes in dynamic loads.

An oscillogram of the kinematic characteristics of the cam mechanism's output link is given in Fig. 9c. Here, the violation of the original symmetric law of motion due excited vibrations is depicted.

Dynamic Analysis of Machines with Two-Way Drive

Figure 10a shows the dynamic model of the two-way drive and its modification (Fig. 10b). Such models are used in machines with long and massive actuators, for example, in some knitting machines. The method of dynamic analysis of this model is similar to that described above, but now Eq. (7) takes the following form:

$$\Gamma = \Gamma_M \Gamma_{J0} \Gamma_u \Gamma_{J*} \Gamma_{c*} \Gamma_{J*} \Gamma_u \Gamma_{J0} \Gamma_M. \tag{14}$$

Figure 11 shows the graphs of the functions β_{jk}, whose minima correspond to the resonance bands (see above). In addition to the previously introduced functions, here, $\beta_{21} = |\tilde{g}_{21}|$. The minimum of this function corresponds to the case when both ends of the system are free. As in the case of a one-way drive (see Fig. 4), the area of change of these functions is selected depending on the position of the input link. Inside this region, a family of curves corresponding to fixed values of the angle φ is shown.

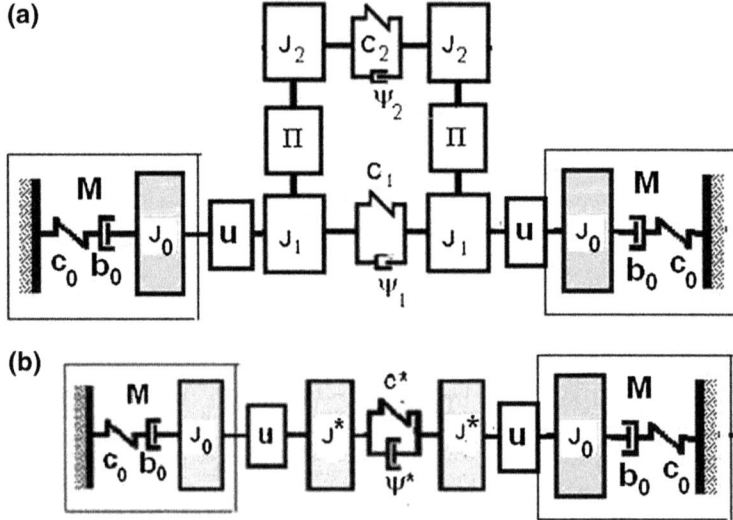

Fig. 10 The dynamic model of a two-way drive (**a**) and its modification (**b**)

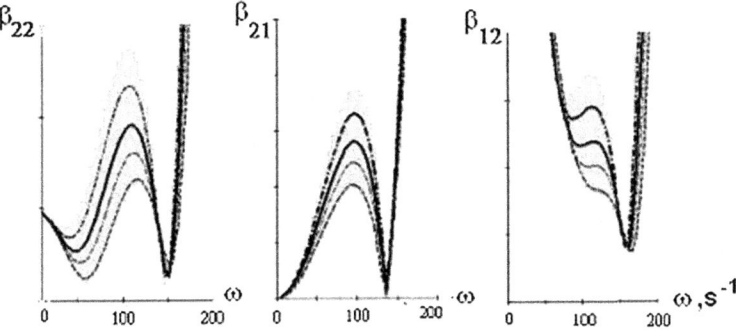

Fig. 11 Graphs of functions β_{jk}

The lowest resonance frequency is clearly manifested in the first minimum of the function β_{22}, which corresponds to a one-way drive (see above). If, as before, we take into account the large "compliance" of the engine ($c_0 \approx 0$), then for two engines, only one positional generalized coordinate remains. The only nonzero variable "eigenvalue" frequency, without allowance for dissipation, is defined as

$$p_*(\varphi) = \sqrt{2c_*(\varphi)/[J_0 + J_*(\varphi)]} \qquad (15)$$

For the initial model (see Fig. 10a), the second frequency corresponds to the minimum of the function β_{12} (pinching of both ends). The difference between the results obtained by formula (15) and the value at the minimum of the function β_{12} is usually no more than 10%.

The question of the efficiency of use of a drive with two engines requires a full-fledged analysis and needs a separate examination. In general, it should be noted that the use of two motors is weakly affected by the characteristics of the oscillatory system, which is due to the large equivalent compliance of motors. In addition, it should be borne in mind that the main resonance at the frequency p_* corresponds to the node of output link vibration, and its ends in the vicinity of this frequency vibrate in antiphase. This can adversely affect the quality of the product. To eliminate this effect, other design solutions are possible when designing a machine. In particular, it may be more efficient to use one engine for a two-way drive with increased rigidity of the connection between the edge elements of the main shaft.

Conclusion

New dynamical models are proposed, on the basis of which the mutual influence of the electric drive and the cyclic machine is investigated. A technique for engineering calculations has been developed and a number of features of this class of dynamical system have been revealed, caused by a large equivalent compliance of the electric motor. A comparative analysis of machines with one-way and two-way drives is carried out. In connection with the problem under consideration, further analysis requires an analysis of the effect of the engine on the excitation of parametric oscillations, as well as on the dynamic effects caused by nonlinear factors (nonlinear couplings, gaps, dissipative forces, etc.).

References

1. Alifov AA, Frolov KV (1985) Interaction of nonlinear oscillatory systems with energy sources, Science, Moscow (in Russian)
2. Veits VL, Kolovskii MZ, Kochura AE (1984) Dynamics of controlled machine aggregates. Nauka, Moscow (in Russian)
3. Kononenko VO (1973) Questions of the theory of the dynamic interaction of a machine and an energy source. Izv. Academy of Sciences of the USSR, MTT. 4:19–30 (in Russian)
4. Banakh LYa.(2016) Methods of decomposition in the study of mechanical systems vibrations, ANO Izhevsk Institute for Computer Research, Izhevsk (in Russian)
5. Pervozvansky AA, Gaizgory VG (1979) Decomposition, aggregation and approximate optimization. Nauka, Moscow (in Russian)
6. Vulfson II, Kolovskii MZ (1968) Non-linear problems of the dynamics of machines. Mashinostroenie, Leningrad (in Russian)
7. Kolovskii MZ (1989) Dynamics of machines. Mashinostroenie, Leningrad (in Russian)
8. Vulfson I (2015) Dynamics of cyclic machines. Springer, Heidelberg, 410 pp (Trans. from Russian, Politechnika, 2013)
9. Vulfson II (2015) Quasistationarity of dynamic regimes in cyclic mechanisms forming rheonomic oscillating systems with lattice structure. J Mach Manuf Reliab 44(4):12–19
10. Vulfson II (1990) Vibrations in machines with cyclic action mechanisms. Machinostroenie, Leningrad (in Russian)
11. Vulfson II (2016) To the problem of the dynamic interdependence of the electric motor and the mechanical drive of the cyclic machine. Theory Mech Mach 32(4):173–182 (in Russian)
12. Vulfson II (2017) Parametric vibrations excitation in cyclic mechanisms. Advances in mechanical engineering. Lecture Notes in Mechanical Engineering. Springer, Berlin, pp 133–143
13. Dresig H, Vulfson II (1989) Dynamik der Mechanismen. Springer, Wien
14. Sorokin ES (1958) Dynamic calculations of the bearing structures. Gosstroyizdat, Moscow, 325 pp (in Russian)
15. Vulfson II (1976) Dynamic analysis of cyclic mechanisms. Machinostroenie, Leningrad (in Russian)

Synthesis of Spherical Four-Links Rotational Pairs in the Solidworks Program

Munir G. Yarullin and Marat R. Faizov

Abstract Modern approaches to the design of spherical four-link rotational pair mechanisms synthesized in the so-called "basic sphere" are analyzed. The approach of the synthesis of spherical mechanisms in the SOLIDWORKS software is offered. Examples of synthesis of various types of spherical mechanism are given.

Keywords Spherical mechanism · Synthesis mechanism · Rotational pair
Types of spherical mechanisms

Introduction

Spherical mechanisms are used quite widely in mechanical engineering. For example, they play the role of manipulators when complex surfaces are being milled [1]. They are compact and, therefore, easily mounted on the hip joints, wrists and at the feet of robots [2]. Spherical mechanisms achieve high angular velocities and angular accelerations due to inherent high strength, low inertia and high dynamics. These mechanisms can be used to orient motion sensors in tracking systems, for example: acting as radar [3], scanning the prototype of the Canterbury mechanism [4], or acting as tracking satellite antennas. The application would enable utilization as a driving mechanism of a solar tracker to follow the sun along the azimuth and altitude [5], as well as for an array of volt cells, which allows for increasing the amount of electricity produced [6, 7]. Moreover, this type of mechanism is used as a device for camera orientation [8, 9], for which a prototype mechanism has been designed, the agile eye.

M. G. Yarullin (✉) · M. R. Faizov
Kazan National Research Technical University named after A.N. Tupolev,
Kazan, Russia
e-mail: Yarullinmg@yahoo.com

M. R. Faizov
e-mail: faizovmarat92@gmail.com

© Springer International Publishing AG 2018
A. N. Evgrafov (ed.), *Advances in Mechanical Engineering*, Lecture Notes
in Mechanical Engineering, https://doi.org/10.1007/978-3-319-72929-9_18

The synthesis of spherical mechanisms has its own characteristic, despite its apparent simplicity. Moreover, an overview of literary sources [10] shows that Russian researchers have unwisely paid little attention to spherical mechanisms.

In this article, we examine an example of the synthesis of spherical four-link rotational pairs in the SOLIDWORKS program.

Structure of the Spherical Four-Link Mechanism

Most simple spherical mechanisms are of the four-link type, containing only one movable rotational pair. The links of the spherical mechanism are necessary for it to perform in the manner of an arc, owing to the fact that they do not touch each other when rotating. Spherical mechanisms are spatial, however, their degree of mobility is determined by the Chebyshev formula (for planar mechanisms). Figure 1 presents a comparison of the different structural schemes of four-links mechanisms: planar joint four-links (a) and spherical joint four-links (b) [11]. For all of the mechanisms specified in this figure, the degree of mobility of the mechanisms is the same:

$$W = 3(m-1) - 2p_1 - 1p_2 = 3(4-1) - 2 \cdot 4 - 1 \cdot 0 = 1, \tag{1}$$

where:

$m = 4$ the number of links,
$p_2 = 0$ the number of dual moving kinematic pairs;
$p_1 = 4$ the number of single moving rotational pairs

The main condition for the formation of a spherical mechanism is that the rotational axes of all pairs must intersect at one point (Fig. 2).

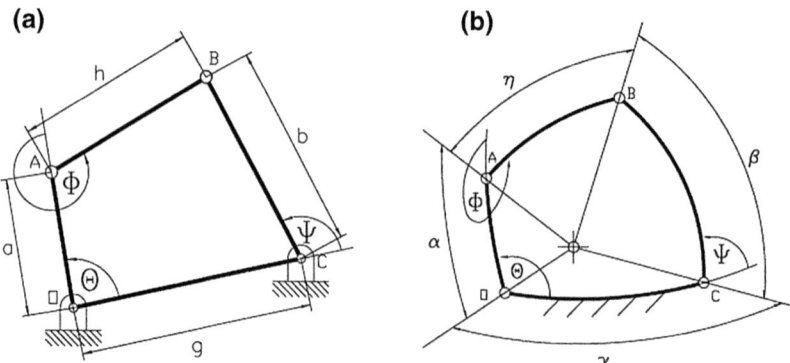

Fig. 1 The structural scheme of four-links: **a** planar joint four-link, **b** spherical four-link [11]

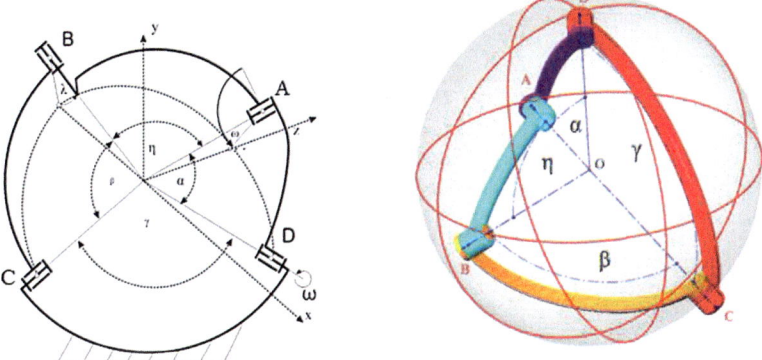

Fig. 2 The structural scheme and the 3D model of a spherical mechanism with rotational pairs only (the axes of all joints intersect at the center of the sphere at the point O)

Synthesis of Spherical Four-Link Mechanisms with Rotational Pairs Only

Different types four-link spherical mechanism exist with only one rotational pair, for example: double-cranks, double-rockers, crank-rockers, etc. For the synthesis of such mechanisms, it is convenient to use the Larochelle condition [12]. Table 1 presents the possible options for types of four-link spherical mechanism. To determine the type of links in a spherical mechanism, Grashof's condition [13] is also used, in the case of a flat four-link: the smallest link is a crank, if the sum of the lengths of the smallest (understood as a link of the minimum length) and any other link is less than the sum of the lengths of the remaining two links [14]. In Fig. 1a, the shortest link's length is a (a = OA), the length of one of the connected links is g, or h, and b is the length of the remaining links of the mechanism. Elementary geometric analysis shows that the condition of full-turn movement of the link of the smallest length relative to the link of length g is the fulfillment of inequality.

$$a + g < h + b. \tag{2}$$

If (g < h) or (g < b), then this inequality will not hold. From this, Grashof's condition is valid in the above formulation.

According to Grashof's condition, there are three basic types of joint four-link:

Double-crank—a mechanism that has two of the same cranks, allowing any of these links to perform a rotational full-reverse motion.

Crank-rocker—this mechanism has one crank and one rocker, and is usually driven by a crank. However, it is possible to set it in motion through the rocker for one particular angle of motion of the rocker, with the crank making a full turn of rotation.

Table 1 Larochelle's basic types of spherical mechanism

No.	Type mechanisms	T_1	T_2	T_3	T_4
1	Crank-rocker	+	+	+	+
2	Rocker-crank	+	–	–	+
3	Double-crank	–	–	+	+
4	Grashof double-rocker	–	+	–	+
5	00+ double-rocker	–	–	–	+
6	0π+ double-rocker	+	+	–	+
7	π0+ double-rocker	+	–	+	+
8	ππ+ double-rocker	–	+	+	+
9	Crank-rocker	–	–	–	–
10	Rocker-crank	–	+	+	–
11	Double-crank	+	+	–	–
12	Grashof double-rocker	+	–	+	–
13	00+ double-rocker	+	+	+	–
14	0π+ double-rocker	–	–	+	–
15	π0+ double-rocker	–	+	–	–
16	ππ+ double-rocker	+	–	–	–

Double-rocker—this mechanism consists of two rockers, a connecting rod and a frame. The mechanism does not allow full-turn movement of any of the links with respect to the frame.

Larochell's paper [12] presents an extended Grashof's theory for types of spherical mechanism. To classify the types of mechanism, the author introduces four parameters: T_1, T_2, T_3, T_4:

$$T_1 = \gamma - \alpha + \eta - \beta; \quad T_3 = \eta + \beta - \gamma - \alpha;$$
$$T_2 = \gamma - \alpha - \eta + \beta; \quad T_4 = 2\pi - \eta - \beta - \gamma - \alpha. \tag{3}$$

The combination of parameters $T_1 \ldots T_4$ determines the type of motion of both the leading and the following links of the four-link spherical mechanism. Larochelle received 84 variants of spherical mechanisms [12]. Table 1 lists its 16 basic types of mechanism.

In the remaining cases (65 types), when any of the parameters T_1, T_2, T_3, T_4 equals zero, a "folding" joint spherical mechanism is obtained. Thus, by calculating the values of the parameters $T_1 \ldots T_4$, we can determine the type of spherical mechanism. On the other hand, the Larochelle approach allows us to set the values of the parameters $T_1 \ldots T_4$ even before the design stage of the mechanism, so that the result is a spherical mechanism of the required type.

Synthesis of a Spherical Mechanism

For synthesis of a spherical mechanism, it is first necessary to calculate the type of spherical mechanism according to the Larochelle method. For example, consider the synthesis of a spherical mechanism with a crank and a rocker wheel (the second type in Table 1). Having determined that the parameters T_1 and T_4 must be greater than zero, and that T_2 and T_3—have a value less than zero, we find the angular values γ, α, η, β. Using the method of selecting the angles for each link (between its hinges), you can fulfill the Larochelle condition. Let us show an example.

From the condition

$$\begin{cases} T_1 = \gamma - \alpha + \eta - \beta \\ T_2 = \gamma - \alpha - \eta + \beta \\ T_3 = \eta + \beta - \gamma - \alpha \\ T_4 = 2\pi - \eta - \beta - \gamma - \alpha \end{cases},$$

we obtain the inequality

$$\begin{cases} \gamma - \alpha + \eta - \beta > 0 \\ \gamma - \alpha - \eta + \beta < 0 \\ \eta + \beta - \gamma - \alpha < 0 \\ 2\pi - \eta - \beta - \gamma - \alpha > 0 \end{cases}.$$

Let $\gamma = 110°, \beta = 90°, \eta = 60°, \alpha = 30°$.

Let us verify the correspondence between the Larochelle requirement (condition 1 in Table 1):

$$\begin{cases} T_1 = \gamma - \alpha + \eta - \beta = 90° - 110° + 60° - 30° = 50° \\ T_2 = \gamma - \alpha - \eta + \beta = 90° - 110° - 60° + 30° = -50° \\ T_3 = \eta + \beta - \gamma - \alpha = 60° + 30° - 90° - 110° = -110° \\ T_4 = 2\pi - \eta - \beta - \gamma - \alpha = 180° - 60° - 30° - 90° - 110° = 70° \end{cases}.$$

After finding the values of the parameters T_1, T_2, T_3 T_4, their signs (positive or negative) are checked according to Table 1. If the conditions coincide with the given type of the selected spherical mechanism, one can proceed to modeling.

Synthesis and 3D modeling [15, 16] of the spherical mechanism was performed with the SOLIDWORKS 2016 software. Synthesis of the spherical four-link mechanism must begin with the construction of the OCD face (Fig. 3).

To create a face, a radius $R = 50$ was specified. Each face has a certain angle of inclination. Figure 3 shows the OCD face with all of the visual dimensions. Each face is saved in a separate file.

They are then assembled, creating all four faces of the spherical mechanism,. To do this, the new file opens a new document, and then selects the "Build" command shown in Fig. 4.

Fig. 3 Example of construction of an OCD face

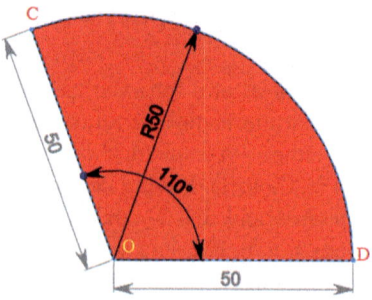

Fig. 4 "Build" document

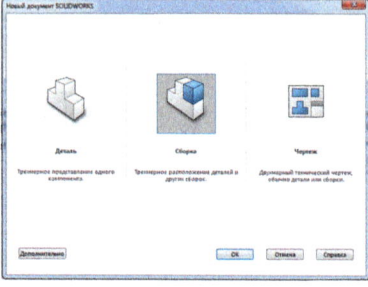

Fig. 5 Toolbar

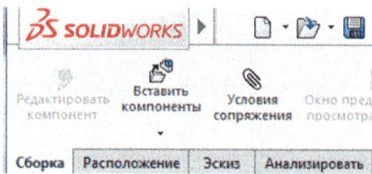

The assembly contains the frequently used commands "Insert Component" and "Pairing Condition". The "Insert Component" command performs the task of inserting parts into the assembly; this toolbar is shown in Fig. 5.

Here is an example of the conjugation of two faces OBC and OCD. Selecting the label "Conjugation conditions" in the toolbar, we first match the points O of the OBC and OCD faces (Fig. 6a).

In this sequence, we mate the other faces (Fig. 6b). On the basis of the face (Fig. 3), a separate link of the spherical mechanism is designed constructively (Fig. 7a) with the same angle γ.

Having constructed the base sphere with the given diameter and centered at the point "O", we place all of the designed links in this sphere, so that the intersecting axes of all of the hinges are located in the center of the sphere (Fig. 7b).

By joining the points A and B consecutively and in pairs (B and C; C and D; D and A), we obtain the links of the spherical four-link mechanism ABCD.

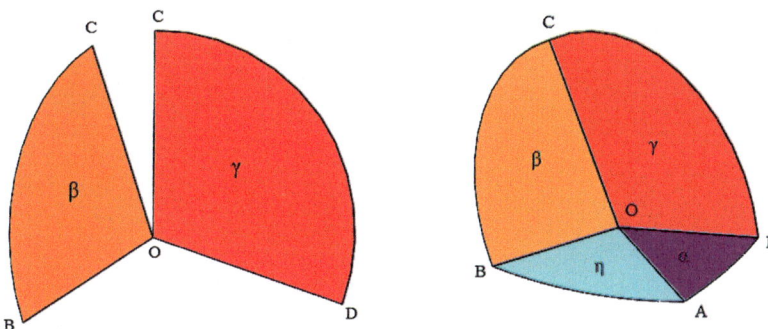

Fig. 6 Example of joining faces by points

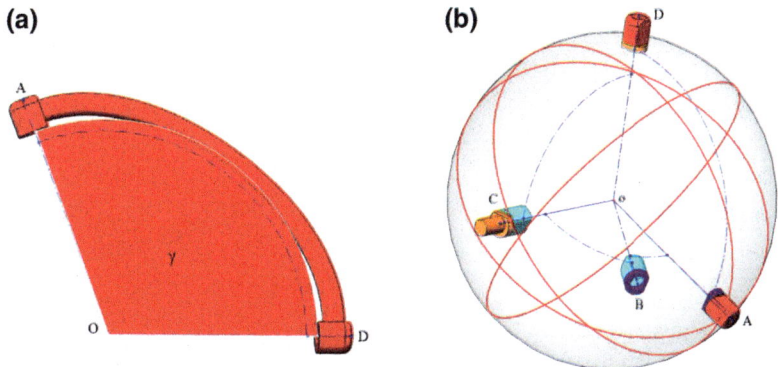

Fig. 7 Schematic representation of links and hinges on the sphere: **a** face designed with a link; **b** the arrangement of all the hinges of the spherical mechanism

Finally, by means of SOLIDWORKS, we synthesize the spherical mechanism, connecting the resulting links of the mechanism successively with hinged bonds (Figs. 8 and 9). As a result, we get a spherical crank-balance mechanism with a rack AD, crank AB, balance wheel CD, and a connecting rod BC, as shown in Fig. 10.

Fig. 8 The two main types of link of the spherical mechanism

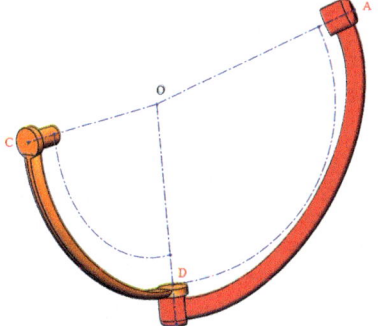

Fig. 9 Joint pivot connection, internal view of the connection

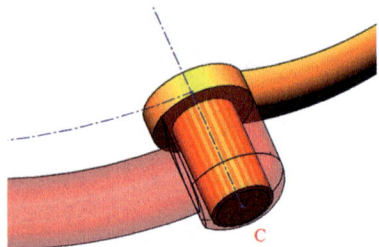

Fig. 10 3D model of the spherical four-link mechanism synthesized in the SOLIDWORKS program

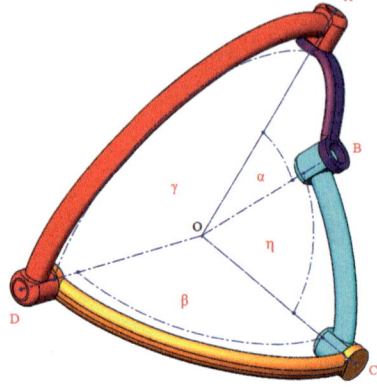

Thus, a 3D model of a spherical four-link mechanism was designed. With this model, you can conduct kinematic and dynamic analyses in the SOLIDWORKS program.

Conclusions

1. The algorithm for designing spherical four-link mechanisms in the SOLIDWORKS system was given.
2. The spherical four-link mechanism of the crank-balance type was synthesized, with the point of intersection of the hinge axes in the center of the base sphere.

References

1. Bonev IA, Chablat D, Wenger P (2006) Working and assembly modes of the agile eye. In: Proceedings 2006 IEEE international conference on robotics and automation, 2006 (ICRA 2006). IEEE, pp 2317–2322
2. Kong X (2010) Forward displacement analysis of a 2-DOF RR-RRR-RRR spherical parallel manipulator. In: 2010 IEEE/ASME international conference on mechatronics and embedded systems and applications (MESA). IEEE, pp 446–451

3. Jiang M, Hu X, Liu L, Yu Y (2012) Study on parallel 2-DOF rotation machanism in radar. Phys Procedia 24:1830–1835
4. Dunlop G, Jones T (1999) Position analysis of a two DOF parallel mechanism the canterbury tracker. Mech Mach Theory 34(4):599–614
5. Rolland L (2012) Manipulators for solar tracking. In: Raad 2012 Proceeding. 21th international workshop on robotics in AlpeAdria-Danube region, 10–13 Sept 2012, vol 1. ESA, Naples, p 93
6. Tonapi SS, Larochelle P (2006) Design of a mirror positioning system to enhance the performance of a PV array. In: Proceedings of the 2006 Florida conference on recent advances in robotics, Citeseer, Miami, 2006
7. Kulkarni S, Tonapi S, Larochelle P, Mitra K (2007) Effect of tracking flat reflector using novel auxiliary drive mechanism on the performance of stationary photovoltaic module. In: ASME international mechanical engineering congress and exposition, vol 6, pp 351–356
8. Gosselin CM, Hamel J-F (1994) The agile eye: a high-performance three-degree-of-freedom camera-orienting device. In: Proceedings, 1994 IEEE international conference on robotics and automation. IEEE, pp 781–786
9. Zhang L-J, Niu Y-W, Li Y-Q, Huang Z (2006) Analysis of the workspace of 2-DOF spherical 5R parallel manipulator. In: Proceedings 2006 IEEE international conference on robotics and automation, 2006 (ICRA 2006). IEEE, pp 1123–1128
10. Kozhevnikov SN, Esipenko Ya I, Raskin Ya M (1976) Mechanisms. Publishing House Mechanical Engineering, Moscow, 784 p (Rus.)
11. Ruth DA, McCarthy JM (1997) SphinxPC: an implementation of four position synthesis for planar and spherical 4R linkages. In: Proceedings of the 1997 ASME design engineering technical conference, Sept 1997
12. Larochelle PM, Dooley JR, Murray AP, McCarthy JM (1993) SPHINX: software for synthesizing spherical 4R mechanisms. In: NSF design and manufacturing systems conference, vol 1, pp 607–611, 3 Jan 1993
13. Frolov KV, Popov SA, Musatov AK (1987) Theory of mechanisms and machines (Ed. Frolov KV). Higher School, Moscow, 496 p (in Russian)
14. Yudin VA, Petrokas LV (1967) Theory of mechanisms and machines. Higher School, Moscow, 528 p (in Russian)
15. Evgrafov AN, Petrov GN (2008) Computer animation of kinematic schemes in excel and mathCAD programs. Theory Mech Mach, T 6. 1(11):71–80 (in Russian)
16. Yarullin MG, Mingazov MR (2014) To the synthesis of spherical mechanisms with rotational pairs. Bull KSTU. A.N. Tupolev, T 70 1:75–80 (in Russian)

Structural Study of a Two-Mobility Five-Link Space Mechanism with a Double Crank

Munir G. Yarullin and Ilnur R. Isyanov

Abstract The present paper aims to analyze the structure of the new two-mobile device of the simulator on the basis of the spatial five-link mechanism. This five-link is created on the basis of the single-moving four-link of Bennett's mechanism. The paper also investigates the method of forming a two-mobility five-link mechanism. It then provides the description of the device with a double crank based on the Bennett mechanism. Furthermore, the present study introduces the definitions of "zero" and "nonzero" links of the mechanism and establishes the working capacity of the new mechanism.

Keywords Bennett's mechanism · Workability · "Zero" and "nonzero" links
Two-mobility · Double crank

Introduction

The interest of design engineers has been significantly piqued by spatial lever mechanisms with only rotational hinges. The initial (fundamental) mechanism for the synthesis of multi-link spatial mechanisms is the Bennett mechanism—a three-dimensional hinged four-link with crossed hinge axes. In 1903, Bennett theoretically confirmed the existence of this mechanism [1], but he could not produce a workable model.

Many foreign and domestic scientists have theoretically investigated Bennett's mechanism by creating mathematical and 3D models [2–5]. However, analysis of both literary and internet sources shows that no one has been able to produce it or, moreover, find a practical application. The problem of manufacturing, as well as application, has been partially solved by the researchers of the Kazan School of

M. G. Yarullin (✉) · I. R. Isyanov
Kazan National Research Technical University named after A.N. Tupolev, Kazan, Russia
e-mail: Yarullinmg@yahoo.com

I. R. Isyanov
e-mail: isyanov1993@mail.ru

© Springer International Publishing AG 2018
A. N. Evgrafov (ed.), *Advances in Mechanical Engineering*, Lecture Notes in Mechanical Engineering, https://doi.org/10.1007/978-3-319-72929-9_19

187

Mechanics, founded by Professors BV Shitikov and PG Mudrov at the end of the 20th century [6]. According to the well-known Somov–Malyshev formula, the mechanism has a mobility of minus two, which means that theoretically, the construction is stationary. However, Bennett's mechanism becomes mobile and workable if the following additional conditions for the existence of the mechanism are fulfilled [7]:

(1) the lengths of the shortest distances of the opposing links are equal and the geometric axes of the opposing kinematic pairs are unfolded with respect to one another at equal angles;
(2) the ends of the shortest distances of the links coincide;
(3) the equality of the lengths and crossings is the same:

$$\frac{l_1}{l_2} = \pm \frac{\sin \alpha_1}{\sin \alpha_2}.$$

Thus, the degree of freedom of the Bennett mechanism is determined by the formula

$$W = 6(n - 1) - 5p_1 + S = 6(4 - 1) - 5 \cdot 4 + 3 = 1, \tag{1}$$

where

n = 4 (the number of Bennett mechanisms links);
p_1 = 4 (the number of single-moving rotational kinematic pairs); and
S = 3 (the number of agreed-upon sizes noted above).

The Kazan School of Mechanics did some work on 3D designing devices based on the Bennett mechanism and their manufacture in metal and plastic. As an example of using Bennett's mechanism as the basic mechanism for synthesizing a two-mobility five-link, consider the device for training the vestibular apparatus.

Bennett's Double Crank Mechanism

A spatial four-link Bennett mechanism is one of the simplest mechanisms with rotational pairs. The mechanism has two crankshafts, a connecting rod, and a rack, and is single-carriage when the above conditions are met. When the driving crank rotates at a constant angular velocity, the driven crank will rotate at a variable speed within one revolution. Figure 1 shows the block diagram of Bennett's mechanism in axonometric. Contour A*B*CD (indicated in red) is a theoretical outline of the structure of the mechanism. The OACD contour is constructed by axonometric images of links, in the form convenient for structural analysis. The structural scheme in this form was considered in the works of the researchers of the Kazan School of Mechanics [8–11].

Since in many devices based on the Bennett mechanism, the connecting rod (the link of the AC in Fig. 1) is used as a working member, it is necessary to ensure the stable operation of this link [12]. To solve this problem, one of the cranks of Bennett's mechanism may be performed doubly, i.e., one theoretical "zero" crank may be converted into two "nonzero" cranks, as shown in Fig. 2. The theoretical "zero" crank A*B* is replaced by two twin "non-zero" cranks O_1A and O_2B.

Fig. 1 Structural diagram of a one-mobile four-link Bennett mechanism

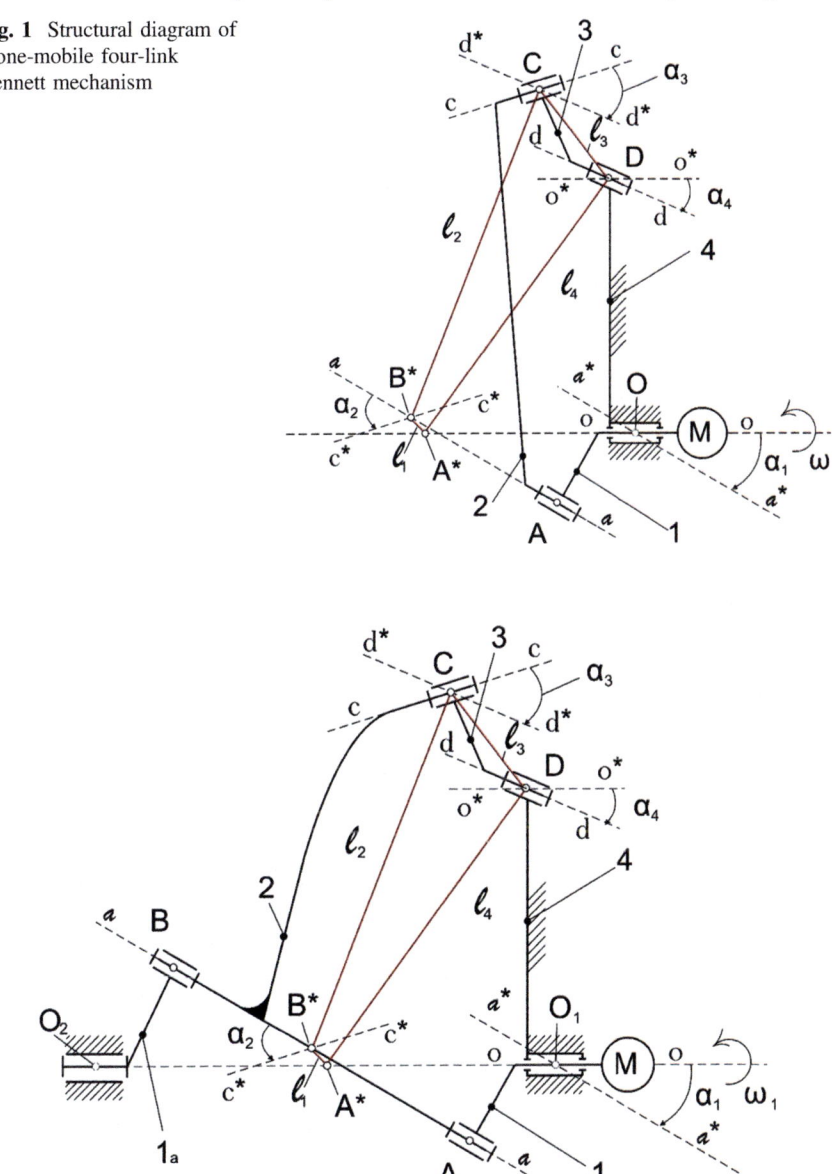

Fig. 2 Structural diagram of the Bennett single-mobile four-link mechanism with a double crank

Though the structural parameters of these two cranks (the shortest distance between the hinge axes—l_1, and the angle of crossing between the hinge axes—α_1) are exactly the same as for the theoretical "zero" crank, it is obvious that the design parameters are different. Thus, in the mechanism shown in Fig. 2, there are three cranks (1, 1a, 3), a connecting rod in the form of a platform (2) and a post (4). All four theoretical links are obtained, since the cranks (1) and (1a) constitute one link.

Two-Mobility Five-Link Mechanism, Created on the Basis of Bennett's One-Link Quadruple

On the basis of Bennett's single-moving four-link mechanism, it is possible to obtain a two-mobility spatial five-link mechanism. The principle of the formation of a two-mobility three-dimensional five-link mechanism is as follows (see Fig. 3). For this, it is necessary, without changing the structural parameters of the Bennett mechanism, to release the rack (4), turning it into a link making rotary movement around the axis of the hinge O_1 (call this link the leading rod in Fig. 3). For the new rack, you need to take the bushing of frame (5), whose axis coincides with the axis "oo" of the leading crank (1) of Fig. 2. The connection of the driving crank (1) and the driving rod (4) must be mounted on the new stand (5), with the possibility of rotation relative to each other, forming two rotational pairs in the hinge O_1.

Fig. 3 Structural diagram of the device for training the vestibular apparatus on the basis of a two-mobility five-link mechanism (synthesized on the basis of the Bennett mechanism with a double crank)

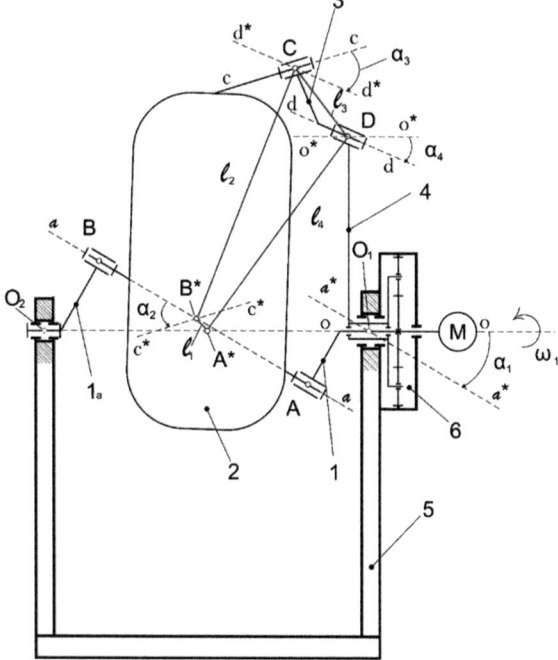

Moreover, the shaft of the driving rod (4) must be made hollow, so that the shaft of the leading crank (1) passes through it, forming a rotational kinematic pair.

Thus, we obtain a two-mobility spatial five-link mechanism. The structural scheme of a two-mobility three-dimensional five-link mechanism is shown in the example of a device for training the vestibular apparatus (Fig. 3). In this scheme, there are five links [a leading double crank (1 and 1a), a connecting rod made in the form of an armchair (2), a driven crank (3), a leading rod (4) and a frame (5)], five single-rod rotary kinematics pairs (hinges A, C, D and two pairs in the hinge O_1). Then, the degree of mobility of the mechanism is

$$W = 6(n-1) - 5p_1 + S = 6(5-1) - 5 \cdot 5 + 3 = 2. \tag{2}$$

The device consists of a frame (5) on which the actuator (6) is mounted and connected to the armchair (2) by means of a basic two-mobility five-link mechanism. The basic mechanism consists of a double crank (1), a leading rod (4), a driven crank (3), and an armchair (2) installed in place of the connecting rod. The double driving crank and the driven crank have the same structural parameters

$$l_1 = l_3, \quad \alpha_1 = \alpha_3, \tag{3}$$

where l_1, l_3 are the lengths of the shortest distance of the driving double crank and the driven crank; and α_1, α_3 are the angles of crossing of the axes of the hinges of the driving double crank and the driven crank.

These links are pivotally connected to the driving rod, armchair and frame. They have the following structural parameters:

$$l_2 = l_4, \quad \alpha_2 = \alpha_4, \tag{4}$$

where l_2, l_4 are the lengths of the shortest distance of the leading rod and armchair; and α_2, α_4 are the angles of crossing of the axes of the hinges of the leading rod and the armchair.

Structurally, in order to simplify the manufacture of links, it is advisable to take the angle of crossing of the geometric axes of the hinges of the driving rod and armchair as being equal to $\alpha_2 = \alpha_4 = 90°$.

By this technique, we have created a number of devices [13–15].

Design Features of the Mechanism

1. Double crank

To ensure the stable operation of the device and improve dynamics, in the basic mechanism, the leading crank of the Bennett mechanism is doubled. Figure 4 shows the scheme for obtaining dual "non-zero" cranks, obtained from the "zero" crank.

Figure 4, the values of the parameters $l_1 = l_3, \alpha_1 = \alpha_3$, are determined from the structure of the Bennett mechanism according to the equation $\frac{l_1}{l_2} = \pm \frac{\sin \alpha_1}{\sin \alpha_2}$.

Parameters h_1, h_1^*, h_2, h_2^* are set constructively, depending on the overall dimensions (for example, the armchair (2) of the simulator presented in Fig. 3) or taking into account the design features of the devices being created.

2. "Zero" and "nonzero" links of the mechanism

Theoretically, using 3D models of Bennett's mechanism, one can determine the mobility of the mechanism and the kinematic and dynamic parameters. A few researchers have done so. When they tried to make the mechanism, it would not rotate, and sometimes it would not assemble. Checking the condition of the assembly, rotation of the mechanism without jamming and establishing links to each other is difficult. The problem of working capacity was very urgent. Therefore, for the production of real models of the Bennett mechanism, especially for the creation of various highly efficient mechatronic devices based on it, it is not enough to have the parameters of its theoretical links. In order for the mechanism to work, and to create new devices and devices on its basis, it is necessary to make the links of the Bennett mechanism in a special way [16]. Therefore, it is necessary to use the so-called "zero" and "nonzero" links of Bennett's mechanism.

First, we give the definitions of the *shortest distance* between the hinge axes of one link (theoretical structural parameter) and *the distance between the hinge centers* of the same link (the constructive parameter of the real link).

Between crossing lines in space, one can always find the shortest distance. The segment of the shortest distance is perpendicular to both crossing lines. This segment is used in the theoretical synthesis of Bennett, and it is called the shortest distance between the crossed axes of the hinges of one link. In Fig. 4, this is the segment A*B*, and on the structural diagrams of Bennett's mechanism, these are the segments A*B*, B*C, CD, DA* (Figs. 1, 2 and 3). In practice, the bodies of links with crossed axes must be displaced by some distances h_1, h_1^*, h_2, h_2^* from this segment A*B* (Fig. 4).

The shortest distance of a link is a segment of the smallest length between the crossed hinge axes, determined by the condition of existence of the mechanism, and is the theoretical structural parameter of the Bennett mechanism.

In addition, there is a concept—*the distance between the centers of the hinges* of one link. For example, in Fig. 4, these are segments AO_1 and BO_2.

Fig. 4 Scheme (3D model) for obtaining dual "non-zero" cranks, obtained from the "zero" crank

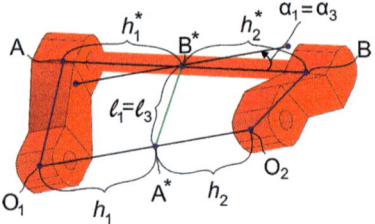

A *"zero" link* is a link whose distance between the centers of the hinges coincides with the shortest distance between the crossed axes of the hinges of this link.

A *"non-zero" link* is a link whose distance between the centers of the hinges does not coincide with the shortest distance between the crossed axes of the hinges of this link and, as a rule, the segment of the shortest distance between the crossed axes is outside the link body, as shown in Fig. 4.

References

1. Bennett GT (1903) A new mechanism. Engineering 76:777–778 (London)
2. Evgrafov AN, Petrov GN (2016) Computer modeling of mechanisms. In: Evgrafova AN, Popovich AA (eds) Modern machine building: science and education: materials of the 5th international scientific-practical conference, Publishing House of Polytechnic University, SPb, pp 203–215. ISSN 2223-0807.1 (in Russian)
3. Evgrafov AN, Petrov GN (2014) Selection of drives of a multimotor mechanism with redundant inputs. In: Radkevich MM, Evgrafova AN (eds) Modern machine building: science and education: materials of the 4th international scientific-practical conference, Publishing House of Polytechnic University, SPb, pp 184–191. ISSN 2223-0807 (in Russian)
4. Nigmatullina FR, Tereshin VA (2014) Kinematic study of the telescope. In: Radkevich MM, Evgrafova AN (eds) Modern machine building: science and education: materials of the 4th international scientific-practical conference, Publishing House of Polytechnic University, SPb, pp 237–246. ISSN 2223-0807
5. Semenov YA, Semenova NS (2014) Static analysis of flat lever mechanisms. In: Radkevich MM, Evgrafova AN (eds) Modern machine building: science and education: materials of the 4th international scientific-practical conference, Publishing House of Polytechnic University, SPb, pp 107–118. ISSN 2223-0807 (in Russian)
6. Yarullin MG, Mudrov AG, Mingazov MR, Galiullin IA (2015) The 1DOF and 2DOF spatial mechanisms with revolute pairs. KNITU-KAI Press, Kazan, 175 p
7. Mudrov PG (1976) Spatial mechanisms with rotational pairs. Kazan Agricultural Institute named after M Gorky, 11 (in Russian)
8. Yarullin MG, Khabibullin FF, Isyanov IR (2016) Nonlinear crushing dynamics in two-degree of freedom disintegrator based on the Bennett's linkage. Vibroeng Procedia 8:477–482. ISSN 2345-0533
9. Yarullin MG, Khabibullin FF (2017) Theoretical and practical conditions of Bennett mechanism workability. In: Advances in mechanical engineering, lecture notes in mechanical engineering, Springer International Publishing AG, Berlin, pp 145–153. ISSN 2195 4356 https://doi.org/10.1007/978-3-319-53363-6
10. Yarullin MG, Isyanov IR, Mudrov AP (2016) Kinematics of a flat two-link five-link lever mechanism. In: Radkevich MM, Evgrafova AN (eds) Contemporary mechanical engineering: science and education: materials of the 5th International scientific-practical conference, Publishing House of Polytechnic University, SPb, pp 297–305. ISSN 2223-0807 (in Russian)
11. Yarullin MG, Mingazov MR (2014) Synthesis of structural modifications of the Bennett mechanism. In: Radkevich MM, Evgrafova AN (eds) Modern machine building: science and education: materials of the 4th international scientific and practical conference, Publishing House of Polytechnic University, no. 4, pp 271–280 (in Russian)
12. Yarullin MG (2002) Intensification of cleaning products in submersible washers based on spatial mechanisms: In: MG Yarullin (ed) Abstract of the thesis of doctor of technical sciences: 05.20.03. MGAU, Moscow, 35 seconds (in Russian)

13. Yarullin MG, Isyanov IR (2015) Two-mobility five-linkage mechanism. Bulletin of KSTU. AN Tupolev, no. 2. ISSN 2078-6255 (in Russian)
14. Pat. 153259 utility model IPC B24B 31/023 the Device for the tumbling dimensionless details, MG Yarullin, RI Isyanov, FF Khabibullin, MR Mingazov// publ. 10.07.2015 bull. No. 19 (in Russian)
15. Pat. 2594302 RU IPC B24B 31/023 the Device for surface dimensionless machining of parts./ Yarullin MG, Isyanov RI, Khabibullin FF, MR Mingazov// publ. 10.07.2015 bull. No. 19 (in Russian)
16. Yarullin MG, Khabibullin FF (2016) To the substantiation of the structural parameters of the mechanisms of the drives of the two-moving disintegrator. In: The XII international scientific and technical conference "VIBRATION-2016". 18.05–20.05, pp 263–269, ISBN 978-5-7681-1116-8 (in Russian)

Modal Analysis of Turbine Blade as One- and Three-Dimensional Body

Tatiana V. Zinovieva and Artem A. Moskalets

Abstract The oscillations of a turbine blade are considered using two models: a naturally twisted rod and a three-dimensional body. The shape of the blade is determined by an array of coordinates of several cross-sections. The boundary value problems for the modes of free oscillations are derived and solved according to the finite difference method. A computer model of the blade as a three-dimensional body is constructed, and its modal analysis is performed in the Ansys program. The natural frequencies found from the two models are compared for various values of the relative length of the blade.

Keywords Turbine blades · Naturally twisted rods · Natural frequencies
Finite element method · Finite difference method

Introduction

Rotating blades of turbo-machines, as well as a turbine disc and a rotor, are exposed to different unsteady forces of the wet steam flow around them [1, 2]. The feature of the disturbing load is its periodicity due to the rotation of the rotor. To ensure strength, the resonance oscillations must not be admitted [3, 4]. The determination of natural frequencies is still of immediate interest. This issue is considered in studies [5–9].

The turbine blade has a small thickness ratio. In the work [10], it is shown that, in this case, the equations of the three-dimensional elasticity split into a one-dimensional problem along the axial coordinate and a two-dimensional one in

T. V. Zinovieva (✉)
Institute of Problems of Mechanical Engineering, Russian Academy of Sciences,
Saint Petersburg, Russia
e-mail: tatiana.zinovieva@gmail.com

A. A. Moskalets
Peter the Great St. Petersburg Polytechnic University, Saint Petersburg, Russia
e-mail: artem.moskalec@gmail.com

© Springer International Publishing AG 2018
A. N. Evgrafov (ed.), *Advances in Mechanical Engineering*, Lecture Notes
in Mechanical Engineering, https://doi.org/10.1007/978-3-319-72929-9_20

195

the cross-section. Therefore, the blade may be considered as an elastic rod, the characteristics of which are given by the geometry of the cross-sections.

On this subject, we already have two papers [11, 12], in which the blade calculation technique is presented for the blade as a naturally twisted rod. However, these studies contain no comparative analysis of results with the blade model as a three-dimensional body. The present study is dedicated to this comparison.

The subject of research is a turbine plant blade, the geometry of which is prescribed by the arrays of Cartesian coordinates of the several cross-sections and an array of axial coordinates of the cross-sections themselves.

We analyze the natural oscillations of the blade as a rod on the basis of the theory of elastic rods. We obtain the system of differential equations and its numerical solution for free oscillations of a naturally twisted rod. Then, to determine numerical results on the basis of analytical calculations, we construct the procedures in the computer mathematics system Mathematica.

The finite element method is implemented in Ansys to determine the natural frequencies of the blade as a three-dimensional body. Finally, we compare the first four natural bending frequencies of the blade determined by two models.

Equations of the Classical Linear Rod Theory

In the first part of the study, the turbine blade is modeled as an elastic rod. Equations of the classical linear rod theory are reliably determined [10]:

$$\mathbf{Q}' + \mathbf{q} = \rho\ddot{\mathbf{u}}, \quad \mathbf{M}' = \mathbf{Q} \times \mathbf{k}, \quad \mathbf{\theta}' = \mathbf{A} \cdot \mathbf{M}, \quad \mathbf{u}' = \mathbf{\theta} \times \mathbf{k}. \tag{1}$$

Rods are considered as material lines whose particles are elementary solid bodies with displacement $\mathbf{u}$ and small rotation vectors $\mathbf{\theta}$. The shape of the rod as a line is given by the dependence of the radius vector $\mathbf{r}(s)$ on the arc coordinate. A prime in Eq. (1) means differentiation with respect to s. Vectors $\mathbf{Q}$, $\mathbf{M}$ determine the force and moment interaction between the rod particles. The load $\mathbf{q}$ is the force distributed per unit length. The vector $\mathbf{k} = \mathbf{r}'$ is the unit tangent vector of the rod axis, and ρ is its mass per unit length. The tensor of second rank $\mathbf{A}$ characterizes the elastic bending and torsional compliances of the rod.

The first equation in system (1) represents the impulse balance (Newton's second law for the rod element). The second equation is the balance of moments. The third equation is the elasticity ratio between the vectors of moment and the bending-torsion strain $\mathbf{\theta}'$. The fourth equation relates the displacement and rotation in the absence of tension and shear, which is characteristic for the classical rod theory.

To solve problem (1) we should first determine the geometric characteristics of the given sections and, using them and interpolation, the elastic properties of the rod.

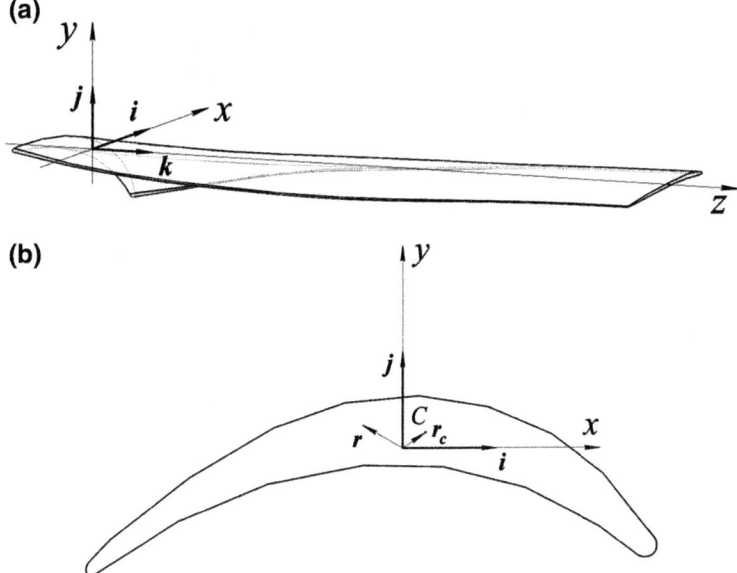

Fig. 1 Blade (a) and its section (b)

We model the rotating blade as a straight, naturally twisted rod of asymmetrical cross-section (Fig. 1), whose shape depends on the axial coordinate.

As shown in [10], the rod's bending compliance is determined from the solution of the Saint-Venant problem for a three-dimensional cylinder:

$$\mathbf{A} = \mathbf{a}_\perp^{-1}, \quad \mathbf{a}_\perp = E\mathbf{I}, \quad \mathbf{I} = -\mathbf{k} \times \int (\mathbf{r} - \mathbf{r}_c)(\mathbf{r} - \mathbf{r}_c)dF \times \mathbf{k}. \tag{2}$$

Here, $\mathbf{a}_\perp$ is the bending stiffness tensor, $\mathbf{I}$ is the central inertia tensor of the cross-section, $\mathbf{k}$ is the unit vector of the rod axis, $\mathbf{r}$ is the position vector of an arbitrary point of the section, $\mathbf{r}_c$ is the position vector of the center of mass, E is the Young modulus, and F is the area. In the introduced coordinate system, the inertia tensor reads as

$$\mathbf{I} = I_x \mathbf{ii} + I_y \mathbf{jj} + I_{xy}(\mathbf{ij} + \mathbf{ji}), \tag{3}$$

where we introduce the symbols of axial $I_x \equiv \int (y - y_c)^2 dF$, $I_y \equiv \int (x - x_c)^2 dF$ and centrifugal $I_{xy} \equiv - \int (x - x_c)(y - y_c)dF$ area moments of inertia.

We note that taking these integrals is not necessary. To determine the geometric characteristics of the cross-sections, it is advisable to seize the opportunities provided by modern CAD/CAE-systems.

In this study, we use the built-in functions of the Ansys preprocessor. The blade's geometry is given by the arrays of Cartesian coordinates $\{x_i, y_i\}$ of its seven cross-sections, and the boundary line and the two-dimensional region itself are easily drawn with these coordinates in the preprocessor. All information necessary for further analysis of the drawn up section can be obtained through the ASUM command: the cross-sectional area, the coordinates of the center of mass and the central moments of inertia. We will use these data later to calculate the compliance of naturally twisted rods. To simplify the process of computation of characteristics of all sections, it is more convenient to write a macro command in the internal Ansys language APDL [13].

For the inertia tensor inversion, let us represent it in the form of a matrix of components in the Cartesian system x, y:

$$I = \begin{pmatrix} I_x & I_{xy} \\ I_{xy} & I_y \end{pmatrix};$$

then, the compliance matrix becomes

$$A = E^{-1} \frac{1}{\left(I_x I_y - I_{xy}^2\right)} \begin{pmatrix} I_y & -I_{xy} \\ -I_{xy} & I_x \end{pmatrix},$$

and the bending compliance tensor of the rod is

$$\mathbf{A} = A_x \mathbf{ii} + A_y \mathbf{jj} + A_{xy}(\mathbf{ij} + \mathbf{ji}), \tag{4}$$

where $A_x = \beta I_y$, $A_y = \beta I_x$, $A_{xy} = -\beta I_{xy}$, $\beta = E^{-1}\left(I_x I_y - I_{xy}^2\right)^{-1}$.

We calculate the compliance components for all seven sections and interpolate them along the coordinate z in the Mathematica computer system [14]. In the same manner, we determine the dependences of mass per unit length $\rho = \rho_v F$, where ρ_v is the bulk density of material. Calculations were made for a blade material with the Young modulus $E = 183$ GPa and bulk density $\rho_v = 8500$ kg/m^3. We show the graphs of the mass per unit length and compliance components for a blade of 1 m length in Fig. 2.

System of Equations for the Blade Bending and Its Numerical Solution

With the computed compliance values (4), Eq. (1) are reduced to the system for the coupled rod bending in two planes x, z and y, z:

$$\begin{aligned} Q_x' &= -q_x + \rho \ddot{u}_x, \quad Q_y' = -q_y + \rho \ddot{u}_y, \quad M_x' = Q_y, \quad M_y' = -Q_x, \\ \theta_x' &= A_x M_x + A_{xy} M_y, \quad \theta_y' = A_{xy} M_x + A_y M_y, \quad u_x' = \theta_y, \quad u_y' = -\theta_x. \end{aligned} \tag{5}$$

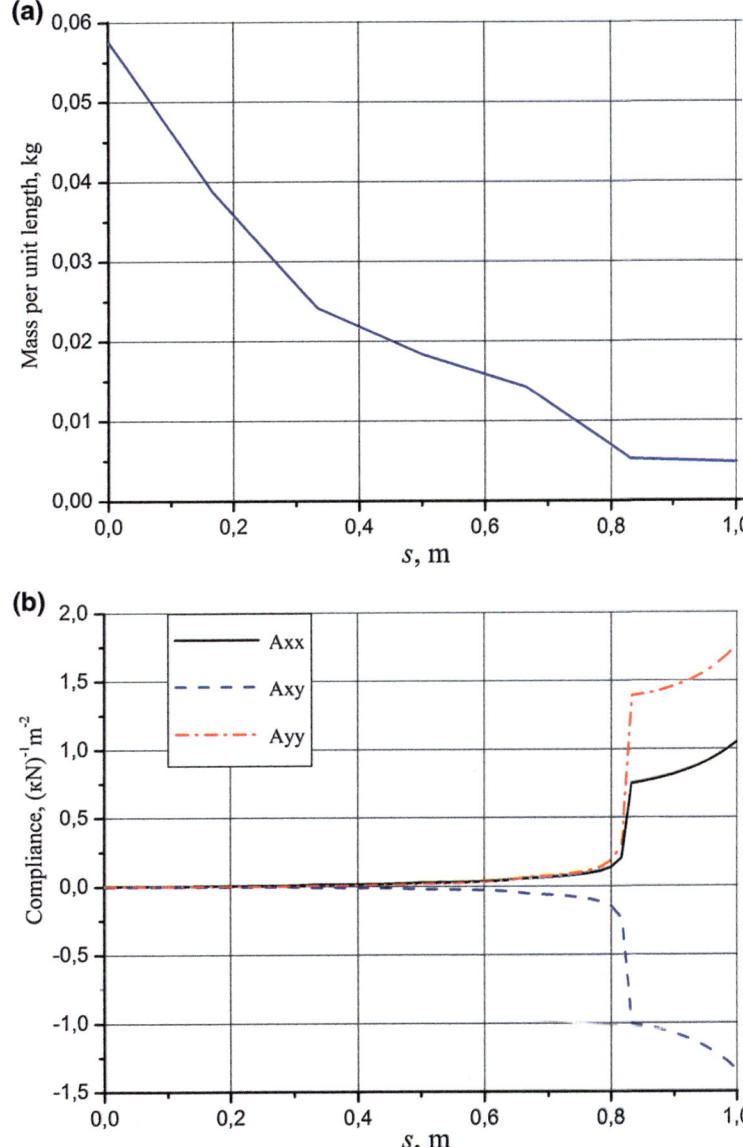

Fig. 2 Dependences of mass per unit length (**a**) and blade bending compliance (**b**) on the axial coordinate

Let us consider the problem of free oscillations of the blade. We will seek the solution of system (5) with no external load in the form proportional to $\sin \omega t$. Hence, we obtain the system for amplitudes:

$$\overline{Q}'_x = -\omega^2 \rho \overline{u}_x, \quad \overline{Q}'_y = -\omega^2 \rho \overline{u}_y, \quad \overline{M}'_x = \overline{Q}_y, \quad \overline{M}'_y = -\overline{Q}_x,$$
$$\overline{\theta}'_x = A_x \overline{M}_x + A_{xy} \overline{M}_y, \quad \overline{\theta}'_y = A_{xy} \overline{M}_x + A_y \overline{M}_y, \quad \overline{u}'_x = \overline{\theta}_y, \quad \overline{u}'_y = -\overline{\theta}_x. \tag{6}$$

We may rewrite these equations in the matrix form

$$Y'(s) = f(Y), \quad Y = \left(\overline{Q}_x, \overline{Q}_y, \overline{M}_x, \overline{M}_y, \overline{\theta}_x, \overline{\theta}_y, \overline{u}_x, \overline{u}_y \right)^T. \tag{7}$$

Equation (7) must be supplemented with the boundary conditions, and we assume that one blade end is fixed and the other end free:

$$z = 0: \quad \overline{\theta}_x = \overline{\theta}_y = 0, \quad \overline{u}_x = \overline{u}_y = 0,$$
$$z = L: \quad \overline{Q}_x = \overline{Q}_y = 0, \quad \overline{M}_x = \overline{M}_y = 0. \tag{8}$$

We solve the boundary value problem (7)–(8) through the finite difference method. The ODE system has the eighth order, as in the case of vibrations of shells of revolution. The solution scheme is described in detail in study [15]. The functions $\overline{Q}_x \ldots \overline{u}_y$ of continuous argument are replaced by the mesh functions $\left(\overline{Q}_x \right)_i \ldots \left(\overline{u}_y \right)_i$, $(i = 0, 1 \ldots, N)$. After approximation of differential equations by difference equations, we arrive at a homogeneous system of linear algebraic equations:

$$B(\omega)\Upsilon = 0, \quad \Upsilon = \left(\left\{ \overline{Q}_x \right\}_i, \ldots \left\{ \overline{u}_y \right\}_i \right)^T, \quad (i = 0, 1 \ldots, N). \tag{9}$$

The values ω, at which system (9) has a nontrivial solution, are the natural frequencies of the rod, which fulfill the following condition:

$$\text{Det } B(\omega) = 0. \tag{10}$$

The roots of Eq. (10) can be found, for example, through the bisection method [16]. The outlined scheme is implemented in the Mathematica package.

Comparison of Results with the Three-Dimensional Blade Model

We drew the three-dimensional blade models in Solidworks. Then, we built the finite element mesh directly in Ansys using 10-node tetrahedral SOLID 187 finite elements with quadratic displacement approximation [17]. The blade material has a modulus of elasticity of $E = 183$ GPa, a Poisson ratio of 0.27 and a bulk density of 8500 kg/m^3.

The modal analysis was carried out with the Lanczos method for four blade models, different in length. Figure 3 shows the dependence of the first four bending

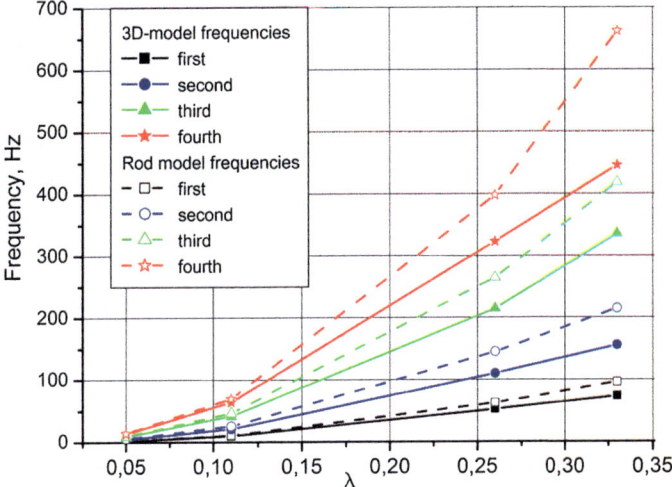

Fig. 3 The blade frequencies calculated according to the rod and three-dimensional models

frequencies of the blade on its thickness ratio $\lambda \equiv d/L$, where d is the root section size and L is the blade length.

The same figure shows the frequencies obtained for the one-dimensional model by formulas given in section "Equations of the Classical Linear Rod Theory". As expected, with an increase in the thickness ratio of the blade, the discrepancy between the two models increases.

The rod model provides overestimated values; this can be explained by the fact that the blade shear and tension compliances were not taken into account in the one-dimensional model.

Figure 4 shows the natural modes of the blade with thickness ratio $\lambda \equiv 0.26$, corresponding to the first four bending frequencies.

For a more accurate comparison of frequencies obtained from a one-dimensional (f_{1D}) and a three-dimensional (f_{3D}) model, we calculate their relative divergence:

$$\Delta \equiv \frac{(f_{1D} - f_{3D})}{f_{1D}} \cdot 100\%,$$

and draw it in Fig. 5.

It turned out that this divergence strongly depends not only on the blade's thickness ratio, but also on the frequency number. Thus, at a thickness λ less than 0.17, both models give the closest values of the fourth frequency ($\Delta < 11\%$), whereas the divergence for the second frequency is 17–20%. This dependence can be explained by the fact that in their natural modes, corresponding to certain blade frequencies, the shear and tension strains are very significant and should be taken into account.

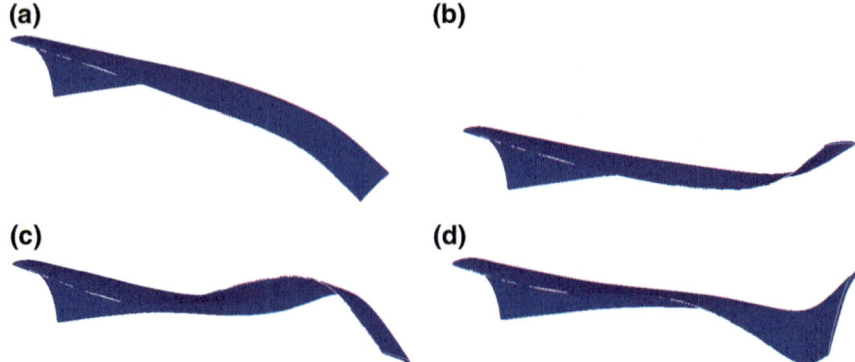

Fig. 4 The blade shapes corresponding to the first (**a**), second (**b**), third (**c**), and fourth (**d**) natural bending frequencies

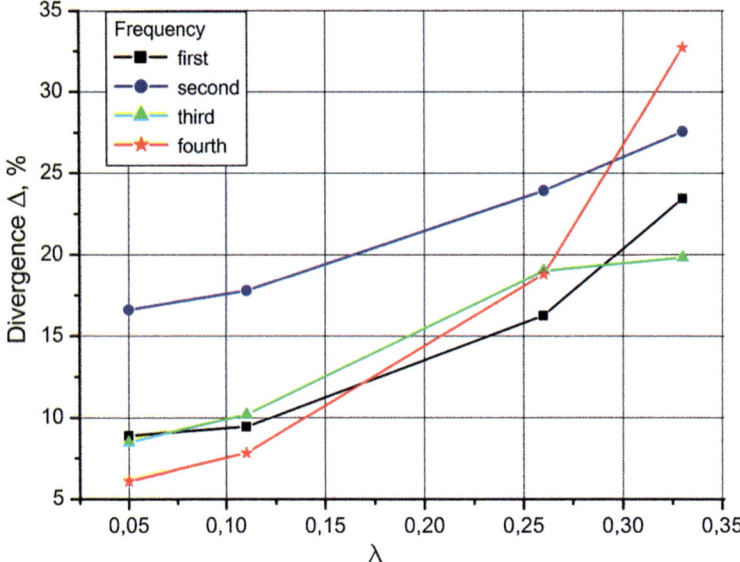

Fig. 5 The divergence between frequencies obtained from the rod and three-dimensional models

Thus, calculations have shown the need to take into account the shear and tension compliance of the blade for a more accurate determination of natural bending frequencies, especially for the second frequency.

Conclusion

In the present study, the modal analysis of a turbine blade as a one-dimensional and a three-dimensional body is carried out. In the first case, a mathematical formulation was used for a straight, naturally twisted rod, its elastic characteristics were determined, and natural frequencies were calculated by solving the ODE system through the finite difference method. Modal analysis of the three-dimensional blade model was carried out through the finite element method using the Ansys program.

The comparison of natural bending frequencies obtained from two models showed that for a blade with length ratio λ less than 0.1, the divergence of results does not exceed 10% for the first, third and fourth frequencies and 17% for the second frequency.

The developed calculation procedure of the turbine blade as a rod can be recommended for use in design practice, because it saves a significant amount of time. However, the procedure should be clarified by taking into account the blade shear and tension compliance.

References

1. Kostyuk AG (1982) Dynamics and strength of turbine engines (Dinamika i prochnost turbomashin). Mashinostroyenie, Moscow, p 264 (in Russian)
2. Bloch HP, Singh MP (2009) Steam turbines: design, application, and re-rating, 2nd edn. McGraw Hill Professional, New York, p 433
3. Levin AV, Borishanskiy KN, Konson ED (1981) Strength and vibration of blades and disks of steam turbines (Prochnost i vibraciya lopatok i diskov parovih turbin). Mashinostroyenie, Leningrad, p 710 (in Russian)
4. Mechanical engineering (1995) Encyclopedia. In: Frolov KV (ed) Dynamics and strength of machines (Book 2). Theory of machines and mechanisms, vol 1–3 (Mashinostroyenie. Dinamika i prochnost mashin. Teoriya mashin i mehanizmov). Mashinostroyenie, Moscow, p 622 (in Russian)
5. Bauer VO et al (1981) Dynamics of aviation gas turbine engines (Dinamika aviacionnyh gazoturbinnyh dvigatelej). Mashinostroyenie, Moscow, p 232 (in Russian)
6. Biderman VL (1980) Theory of mechanical vibrations (Teoriya mekhanicheskih kolebanij). Vyisshaya shkola, Moscow, p 408 (in Russian)
7. Vibrations in technics (1980) Handbook. In: Dimetberg FM, Kolesnikov KS (ed) Vibrations of machines, constructions and their elements, vol 3 (Vibracii v tekhnike. Vibracii mashin, konsrukcij i ih elementov). Mashinostroyenie, Moscow, p 544 (in Russian)
8. Vorobiev YS, Shulzhenko NG (1978) Investigation of oscillations of turbine machines elements systems (Issledovanie kolebanij sistem ehlementov turboagregatov). Naukova Dumka, Kiev, p 134 (in Russian)
9. Zhiritskiy GS, Strunkin VA (1968) Design and strength calculation of parts of steam and gas turbines (Konstrukciya i raschet na prochnost detalej parovyh i gazovyh turbin). Mashinostroyenie, Moscow, p 520 (in Russian)
10. Eliseev VV (2003) Mechanics of elastic bodies (Mekhanika uprugih tel). Saint-Petersburg State Polytechn. University Publishing House, Saint-Petersburg, p 336 (in Russian)

11. Eliseev VV, Moskalets AA, Oborin EA (2016) One-dimensional models in turbine blades dynamics. Advances in mechanical engineering, Lecture notes in mechanical engineering. Springer, Berlin, pp 93–104. https://doi.org/10.1007/978-3-319-29579-4
12. Eliseev VV, Moskalets AA (2014) Vibrations of turbine blades as naturally twisted rods (Kolebaniya turbinnyh lopatok kak estestvenno zakruchennyh sterzhnej). In: Materials of 4th international scientific and practical conference "Modern Engineering: Science and Education". Saint-Petersburg Polytechn. University Publishing House, Saint-Petersburg, pp 344–350 (in Russian)
13. Eliseev KV, Zinovieva TV (2008) Computational practical course in modern CAE-systems (Vychislitelnyj praktikum v sovremennyh CAE-sistemah). Saint-Petersburg State Polytechn. University Publishing House, Saint-Petersburg, p 112 (in Russian)
14. Borwein JM, Skerritt MB (2012) An introduction to modern mathematical computing: with Mathematica, vol XVI. Springer, Berlin, p 224
15. Zinovieva TV (2017) Calculation of shells of revolution with arbitrary meridian oscillation. Advances in mechanical engineering, Lecture notes in mechanical engineering. Springer, Berlin, pp 165–176. https://doi.org/10.1007/978-3-319-53363-6_17
16. Chapra SC, Canale RP (2014) Numerical methods for engineers. McGraw-Hill Education, New York, p 992
17. ANSYS Inc. PDF Documentation for Release 15.0

Author Index

© Springer International Publishing AG 2018
A. N. Evgrafov (ed.), *Advances in Mechanical Engineering*, Lecture Notes
in Mechanical Engineering, https://doi.org/10.1007/978-3-319-72929-9

Printed by Printforce, the Netherlands